The Essential Guide to Mobile Business

ISBN 0-13-093819-X

Essential Guide Series

**THE ESSENTIAL GUIDE TO
DATA WAREHOUSING**
Agosta

**THE ESSENTIAL GUIDE TO WEB STRATEGY
FOR ENTREPRENEURS**
Bergman

**THE ESSENTIAL GUIDE TO THE BUSINESS
OF US WIRELESS COMMUNICATIONS**
Burnham

**THE ESSENTIAL GUIDE TO TELECOMMUNICATIONS,
SECOND EDITION**
Dodd

**THE ESSENTIAL GUIDE TO WIRELESS COMMUNICATIONS
APPLICATIONS: FROM CELLULAR SYSTEMS TO WAP
AND M-COMMERCE**
Dornan

THE ESSENTIAL GUIDE TO NETWORKING
Keogh

**THE ESSENTIAL GUIDE TO COMPUTER DATA STORAGE:
FROM FLOPPY TO DVD**
Khurshudov

**THE ESSENTIAL GUIDE TO DIGITAL SET-TOP BOXES
AND INTERACTIVE TV**
O'Driscoll

**THE ESSENTIAL GUIDE TO HOME NETWORKING
TECHNOLOGIES**
O'Driscoll

**THE ESSENTIAL GUIDE TO APPLICATION SERVICE
PROVIDERS**
Toigo

**THE ESSENTIAL GUIDE TO KNOWLEDGE MANAGEMENT:
E-BUSINESS AND CRM APPLICATIONS**
Tiwana

THE ESSENTIAL GUIDE TO MOBILE BUSINESS
Vos & de Klein

**THE ESSENTIAL GUIDE TO COMPUTING:
THE STORY OF INFORMATION TECHNOLOGY**
Walters

THE ESSENTIAL GUIDE TO RF AND WIRELESS
Weisman

The Essential Guide to Mobile Business

Ingrid Vos
Pieter de Klein

Prentice Hall PTR, Upper Saddle River, NJ 07458
www.phptr.com

Library of Congress Cataloging-in-Publication Data available

Editorial/Production Supervision: *MetroVoice Publishing Services*
Acquisitions Editor: *Bernard Goodwin*
Editorial Assistant: *Michelle Vincenti*
Manufacturing Manager: *Alexis Heydt-Long*
Art Director: *Gail Cocker-Bogusz*
Interior Series Design: *Meg Van Arsdale*
Cover Design: *Talar Agasyan-Boorujy*
Cover Design Direction: *Jerry Votta*

© 2002 Prentice Hall PTR
Prentice-Hall, Inc.
Upper Saddle River, NJ 07458

All rights reserved. No part of this book may be reproduced, in any form or by any means, without permission in writing from the publisher.

The publisher offers discounts on this book when ordered in bulk quantities.
For more information, contact
Corporate Sales Department,
Prentice Hall PTR
One Lake Street
Upper Saddle River, NJ 07458
Phone: 800-382-3419; Fax: 201-236-7141
Email (Internet): corpsales@prenhall.com

Printed in the United States of America

10 9 8 7 6 5 4 3 2 1

ISBN 0-13-093819-X

Pearson Education Ltd.
Pearson Education Australia PTY Ltd.
Pearson Education Singapore, Pte. Ltd.
Pearson Education North Asia Ltd.
Pearson Education Canada, Ltd.
Pearson Educación de Mexico, S.A. de C.V.
Pearson Education—Japan
Pearson Education Malaysia, Pte. Ltd.
Pearson Education, Upper Saddle River, New Jersey

Contents

Preface xv

    Structure of the Book xvi

1 WAP and the Future 1

    WAP 2

        Specially Designed for Mobile Communication 2
        The WAP Building Blocks 6
        Telephony Services within WAP 15
        Content Billing 16
        Security 17
        SIM Application Toolkit 19
        WAP and USSD 21
        WAP on the SIM Card 21

    Cellular Network Standards 22

        First-Generation Mobile Networks 22
        Second-Generation Mobile Networks 23
        WAP and 2G Mobile Networks 24
        2.5-Generation Mobile Networks 25
        Third-Generation Mobile Networks 28

    New Possibilities 35

        Positioning 35
        Synchronization 38
        Speech Technology 38
        Bluetooth 39

Always and Everywhere   41

    Standards   42

    Sharpened Service Offerings   42

    Complementary Technologies and Possibilities   42

2 Similarities and Differences between WAP and the Internet   45

WAP versus the Internet for the User   46

    Differences between Mobile Phones and the PC   47

    Voice Recognition   53

    Telephony Services   54

    Availability   54

    Threshold of Purchase   54

    User Threshold   55

    Way of Use   55

    Number of Users per Connection   56

    Perception of Security   56

    Methods of Payment   57

    Price Sensitivity   57

    Integration of WAP and the Internet   58

    Unified Messaging   58

WAP versus the Internet for the Service Provider   61

    Characteristics of a Mobile Phone   61

    Speed   62

    Up-to-Date Information   62

    Integration with Telephony Services   63

    Availability   64

    Barrier to Purchase   65

Contents

User Threshold   65

Personal Use   66

Security   67

Invoicing   67

Price Perception and Pricing   68

Integration Possibilities   69

WAP versus the Internet for the Service Developer   70

Characteristics of a Mobile Phone   71

Speed   72

Integration with Speech   72

Purchase Threshold   72

User Threshold   73

Personal Use   74

Security   74

Billing Possibilities   75

Integration Possibilities   75

WAP and Corporate Networks   77

3 The Rapid Development of Mobile Communications   81

Mobile Communication   82

Different Roles in Mobile Communication   86

Mobile Operators   86

Service Providers   92

Network Suppliers   94

Mobile Telephone Suppliers   94

Mobile Phone Usage   96

Mobile Company Networks   98

4 The Internet   101

Different Roles in Service Provision   103

Internet Access and Service Providers   103

Software Suppliers   105

Hardware Suppliers   106

Network Suppliers   106

Portals   106

Content Providers, Merchants, and Advertisers   107

Internet Usage   110

Company Networks   112

5 Mobile Internet   113

Mobile Internet outside Europe   117

i-mode in Japan   118

Other Developments in Asia   121

Mobile Internet in the United States   121

Different Roles with Mobile Internet   122

Mobile Phone Suppliers   122

The Fight for Market Share for Hybrid Devices   124

Browser Suppliers   126

Software Tools Suppliers   128

Gateway Suppliers   128

Mobile Operators   129

Mobile Internet Service Provider (M-ISP)   131

Mobile Portals   133

Transaction Providers   137

Content Providers   138

Merchants   139

Advertisers   140

Conflicts in the Value Chain   143

Mobile Access   143
Advertising   144
Location-Based Services   144
Transactions   144
Customer Ownership   145
Brands   146

6 Mobile Internet: Get on Board Now, or Wait?   149

Trends in Society   151

Downaging   151
Clanning   153
Fantasy Adventures   154
Egonomics   156
99 Lives   157
The Vigilante Consumer   159
Female Think   160
Cocooning   161
Small Indulgences   163
Cashing Out   164
Icon Toppling   165

Start Now or Wait   166

Mobile Internet in Business-to-Business Relationships   168

Mobile Internet as Trend Accelerator   169

7 Opportunities and Threats of Mobile Internet for Your Business   171

Customers   173

Transparent Information   173

Improved Customer Service   174

Payment   175

Impulsive Actions   176

Competition   177

Ease of Use   177

"In-Store" Competition   178

Additional Communication Channels   178

New Entrants to Market   179

Borders Are Less Relevant   179

The Pioneer's Advantage   180

Brand Power   180

Physical versus Virtual Presence   181

Deep Pockets   182

Knowledge   182

Substitutes   183

A New Communication Channel   183

Information   184

Distribution   185

Entertainment   186

Transactions   186

Suppliers   188

Direct Sales   188

Rise of Vending Machines   188

Using Extranets   189

8 WAP in the Real World   191

CMG—ICT Services Group   192

Emerce—E-Business Publisher   193

Emerce and WAP   194

iMedia—Multimedia Design Agency   195

iMedia and WAP   195

De Telegraaf—Newspaper Publisher   196

Business Model and Objectives   197

Developing WAP Services   198

Advice to Companies Considering WAP   198

Scoot—Finder Service   198

Atos Origin—IT Services Company   199

Lotus—Software Company   201

Market Leader   201

@info Info   203

Bruna—Retail Chain   203

Bibit—Payment Service Provider   205

123internet—Content Organizer   206

123internet and WAP   207

AtoBe—Wireless Application Service Provider   208

Finphone—Financial Information Provider   210

Siennax—Application Service Provider   211

Siennax and WAP   212

Sky Radio—Radio Station   212

Twigger—Worldwide Email Access 214

WAPDrive—WAP Portal 215

WAPDrive Service Categories 216

XS4ALL—Internet Service Provider 218

9 Five Steps to a Successful WAP Site 221

Step 1: A Clear Objective 222

Step 2: The Marketing Mix 225

Product 225

Price 232

Place 233

Promotion 235

Personnel 238

Critical Success Factors for the Marketing Mix 239

Step 3: Financial Analysis 240

Costs 240

Revenue 243

Step 4: Plan of Action 245

Step 5: Control and Reporting 245

10 A Look into the Mobile Future 247

Appendix A: Examples and Tips to Build a WAP Site — 253

Example 1: A Flash Screen with Logo — 255

Example 2: Links to Other Pages — 256

Example 3: The Use of a Parameter — 257

Example 4: Selection from Options — 259

Appendix B: List of Interesting URLs — 261

Information about WAP — 261

The WAP Forum — 262

Bluetooth — 262

Make Your Own WAP Home Page — 262

WAP Developers Toolkits — 262

Converter for Bitmaps — 262

WAP Suppliers — 262

PDAs — 263

Mobile Operators — 263

Content — 263

Consultants — 264

Abbreviations — 265
Index — 269
About the Authors — 279

Preface

The mobile Internet has attracted a lot of media attention. From Japan to the United States, the promise of Internet access via your mobile phone seemed a revolutionary new step. After the hype there was also a sense of deception once it was offered to the public in most countries around the world. Was this what we had been dreaming about? Except for the amazing success of i-mode in Japan, the first WAP-based services often didn't attract the number of customers predicted in the first year.

What happened? In most countries, WAP was hyped by vendors of mobile phones or network equipment as the mobile Internet, showing images on mobile phones, which will not be possible for years to come. So when people finally got hold of a mobile phone capable of using WAP (availability was very limited the first year), they were disappointed that the services on their screen were not even close to the services shown in all the ads.

What's next? After this initial disappointment, people now have a more realistic expectation of what mobile access to the Internet can mean to them. Mobile operators and service providers create knowledge to improve services and customer satisfaction. Most mobile phones sold today are equipped with a WAP browser or (c)HTML browser, creating more opportunities for successful implementation and marketing of mobile Internet services.

But how? Don't make the mistake of thinking that the mobile Internet is equal to the "fixed" Internet, but mobile. Mobile phones have their own particular usage patterns and characteristics that need to be taken into account when designing and marketing a mobile Internet service.

Why this book? We believe that the mobile phone will become part of almost everybody's life and that people will want to use it for more than just talking with friends, family, or business relations. We predict that the mobile phone will grow from a voice communications device into a multipurpose terminal allowing people to communicate, exchange messages and email, use it as their information terminal and guide, their personal shopper, and even as their wallet. In this book, we describe both the technical developments and the societal trends that lead us to believe this. We explain what WAP and i-mode are, which markets will emerge because of the mobile Internet, and the opportunities this offers you and your company.

Who should read this book? Each manager interested in the developments of new channels to his or her customers or better communications with his or her sales force and other employees in the field, should read this book. Although the book starts with an overview of the technical developments, this is by no means a technical book. We looked at WAP and the mobile Internet from a business and marketing point of view. We tried to offer informative but easy-to-read material to a broad group of people interested in new developments—people wondering how to benefit from those developments.

STRUCTURE OF THE BOOK

Chapter 1 explains WAP and other technical developments like i-mode, GPRS, UMTS, SMS, location determination, and Bluetooth. We explain the abbreviations and, more important, describe which possibilities they offer for a user.

WAP and the mobile Internet are often named in one breath. Chapter 2 shows the similarities and differences between WAP and the use of the Internet via the PC. We show this from the perspective of the user, the service provider, and the service developer.

In Chapters 3, 4, and 5, we explore the markets for mobile telephony, the Internet, and mobile Internet. The developments and roles in each of these markets are explained and the use of mobile telephony and the Internet is described. At the end of Chapter 5, the playing field for WAP and the mobile Internet is clear from both a functional and a business perspective and we turn our perspective to the reader's situation. Which services could you offer your customers using the mobile Internet? In Chapter 6, we look at the most important trends in society you can react to with mobile Internet services. In the next chapter, we look at your playing field. What will customers expect from you in the future? Which movements will suppliers make? What will your competitors do? Will new players enter your market? Which substitution effects can you expect? In Chapter 8, we visit different companies that already have experience with WAP. What activities do they deploy with WAP? What is their business model? What are their experiences with the introduction of a WAP service?

In Chapter 9, we offer a five-step approach to a successful WAP site. Chapter 10 gives a view of a day in the mobile future. Of course it's up to you to think and act toward an even better mobile future. We hope this book contributes to your mobile Internet business success.

We were, as marketers, responsible for the introduction of the first commercial WAP service in Europe: @info. This experience along with a lot of additional research has enabled us to write this book.

1 WAP and the Future

In this chapter...

- WAP   2
- Cellular Network Standards   22
- New Possibilities   35
- Always and Everywhere   41

This chapter concentrates on new techniques such as WAP and GPRS. Benefits of these techniques and the possibilities they offer to current and future users are described. Thanks to these techniques, everyone will be able to access the Internet from mobile telephones, organizers, or other mobile equipment. Via Internet, a wide range of other services will be accessible. Also, related technologies like EDGE, UMTS Bluetooth, and mobile positioning are explained.

WAP

Specially Designed for Mobile Communication

Wireless Application Protocol (WAP) is a standardized protocol that enables an application to be set up between a mobile telephone and a server. This enables Internet access, as well as use of other applications. The mobile user (Figure 1.1) can access information and applications on the Internet or on his or her company's Intranet.

This was already possible before the introduction of WAP. A user who has access to a laptop or organizer can connect to Internet via mobile telephone. The user calls into an Internet service provider via a mobile phone in exactly the same way that

Figure 1.1
Example of a WAP-enabled phone: the Ericsson R320.

he or she would do via a fixed telephone line. The user's interface is identical, but the speed is considerably slower. The bandwidth of a GSM connection is 9.6 kbit per second compared to the 56.6 or 64 kbit per second maximum provided by a fixed PSTN or ISDN telephone network. The GSM connection will be disconnected more often and give more transmission faults than a connection via a fixed telephone network because it is a radio connection.

The Internet is a simple and efficient method of delivering services to millions of PC users. WAP was specially developed to make the convenience of the Internet available to mobile users without the need for a laptop. This means that WAP takes into account the following limitations of the mobile terminal and the mobile connection:

- *Low bandwidth.* Mobile networks have a lower bandwidth than fixed networks. This leads to lower performance. WAP minimizes traffic over the mobile network, enabling better performance. The WAP protocol eliminates unnecessary information, enabling only about half the quantity of data that is needed by a standard HTTP/TCP/IP stack to deliver the same content.
- *Long delays.* Mobile networks have longer delays than fixed networks due to the lower data speed. For the user this means a long wait before a reaction to his or her action. WAP minimizes the number of question-and-answer sessions between the mobile appliance and the WAP device.
- *Poor connection stability.* Mobile networks can be unavailable to the user for shorter or longer periods, for example, due to limited capacity or breakdown in radio contact. WAP minimizes the effects of signal failure by maintaining the logical session intact. Also, specific lost data can be resent on a selective basis.
- *Limited screen size.* Mobile terminals (even PDAs) have a small screen compared to a desktop. The size of the human hand will always limit screen size, even with the arrival of slightly bigger handhelds. WAP takes account of the small screen by breaking the total interaction with the user up into small parts (the so-called cards) that can be shown on small screens. Examples of interactions are text screens, lists of options, input fields, or combinations of these.
- *Limited input possibilities.* Mobile terminals have a small keyboard for data input. This is clearly much more difficult to use than the QWERTY keyboard on a PC. Text as well as numbers can be input with WAP, but input will generally be limited.

- *Limited memory and processor capacity.* Mobile terminals have very limited memory and processing speed compared to PCs. This is true for working memory and space for programs that the manufacturer puts in the terminal. WAP needs little memory or processing speed.
- *Limited battery capacity.* Access to mobile services increases the use of the radio interface and with it the consumption of power. WAP restricts the use of bandwidth and thus also the battery consumption.

The WAP standard has been developed within the WAP Forum and it is being further developed and extended. WAP is an initiative from Unwired Planet (now Openwave), Nokia, Ericsson, and Motorola. By 2001 more than 600 companies were members of the WAP Forum. Among these companies are mobile operators with more than 100 million subscribers, and suppliers of mobile terminals and Personal Digital Assistants representing over 90 percent of the global handset market. WAP Forum also represents all other relevant parties like vendors of SIM cards, hardware and software, and also banks, Internet portals, car suppliers, Internet service providers, and consultancy companies. The WAP Forum (*www.wapforum.org*) develops new WAP standards specifications and certifies WAP products to improve interoperability of the many different WAP-related products.

OPENWAVE, THE FOUNDER OF WAP

When the broadband hype broke out in Silicon Valley at the end of 1994, Alain Rossmann, the founder of what is now Openwave, set himself the goal of making the Internet accessible via mobile telephones. In spite of all the limitations of a small device, Rossmann saw the value of Internet on the cellular: You always have it with you. He also foresaw a mass market. In 1996, AT&T was the first company to step into the mobile Internet world with the introduction of PocketNet. The product did not take off in the mass market, mainly because there were only a few types of large, heavy devices available. In the business market, numerous applications developed for transportation and logistics are still in use.

In 1997, in order to break into the fast-growing global mass market for mobile telephony, Openwave (then called Unwired Planet) decided to share its technological lead with dominant mobile telecom suppliers Ericsson, Nokia, and Motorola with the aim of making its technology the de facto standard. The WAP Forum was born. This "new economy" decision was certainly good for Unwired Planet. The WAP Forum developed into the industrial platform for mobile Internet. Unwired Planet changed its name to Phone.com. Phone.com itself has a developers forum where more than 100,000 developers are registered. The Phone.com microbrowser has been licensed to more than 20 mobile telephone suppliers and is used by Motorola, Alcatel, Panasonic, Siemens, and Samsung, among others. The French Vodafone daughter, SFR, was the first European mobile operator to take the plunge. In March 1999, E.medi@ saw the light, as part of a mobile telephone package for small businesses. E.medi@ was based on Phone.com's own HDML programming language, the basis for WAP's WML.

Phone.com's application for a stock market offering in June 1999 gave financial room for expansion. Sales offices have now been opened in all important mobile markets. Also, Phone.com has carried out a number of acquisitions. They have bought Apion to serve the European operator market, @motion to add mobile speech recognition, Paragon for synchronization products for synchronization of PDAs and mobile telephones with the most important organizer platforms, and Onebox for unified messaging. These acquisitions have enabled Phone.com to expand its gateway product with personalization modules, telephony applications, and Intranet applications. Also, Phone.com has released a separate synchronization product under the name FoneSync Essentials (*www.fonesync.com*). This product enables synchronization of the most important organizer platforms (e.g., Microsoft, Lotus) with a whole range of types of mobile telephones and PDAs.

In November 2000, Phone.com and Software.com, a developer of Internet infrastructure software for communications service providers, merged, adding a broad range of messaging products to the Phone.com portfolio. The merged company was called Openwave and it currently serves mobile operators in Asia, Europe, and the U.S. In the Japanese market—the biggest and most advanced mobile Internet market at the moment—Openwave has IDO and KDDI as customers. With i-mode, Japanese market leader NTT DoCoMo has developed its own WAP variant (see the section "Japanese Competition for WAP?"). Openwave claims to be the worldwide leader of open Internet-based communication infrastructure software and applications with more than 80 mobile operator customers like Vodafone Mannesmann (Germany), Telecom Italia, Sprint, and British Telecom.

The first commercial WAP services were based on the WAP 1.1 standard, issued in June 1999. The first commercial services were launched at the end of 1999 (@info from KPN Mobile premiered on November 25, 1999). Prior to WAP 1.1, a number of mobile operators carried out a pilot with WAP 1.0. WAP 1.1 deviates somewhat from the WAP 1.0 standard and equipment based on WAP 1.0 cannot be used for services that are based on WAP 1.1. The Siemens S25 handset used a WAP 1.0 browser that was not compatible with any WAP service.

The WAP 1.2 specifications were approved in June 2000. The first commercial services based on WAP 1.2 were introduced early 2001. WAP 1.2 is backward-compatible with WAP 1.1. This means that WAP 1.1 handsets work with WAP 1.2-based services. WAP 1.2 provided several technical enhancements of WAP 1.1 functionality and a range of new functionality—SMS push services, end-to-end security using the SIM card (or another smart card), and integration of voice telephony services. The characteristics of the new functionality will be explained later.

WAP 2.0—released in Summer 2001—incorporates TCP/IP and xHTML into the WAP standard, offering programmers the facility to develop an application for both fixed and wireless Internet at once. WAP 2.0 also anticipates the upcoming faster networks and services, solving earlier stumbling blocks like security, personalization, and provisioning. WAP 2.0 should also be backward-compatible with WAP 1.x.

The WAP Building Blocks

In many respects WAP resembles the Internet. An example is the manner in which interaction takes place. With WAP the user also has a browser at his or her disposal, enabling a request for an Internet address to be entered. This request is transferred to a WAP gateway via a mobile connection. This gateway sends the information request on to the WAP server. The server sends the required information back via the gateway. The gateway sends it back to the mobile phone over the mobile connection.

The WAP building blocks (Figure 1.2) are:

- the WAP client (the browser in the mobile telephone)
- the WAP gateway
- the server
- supporting services
- networks
- protocols

WAP

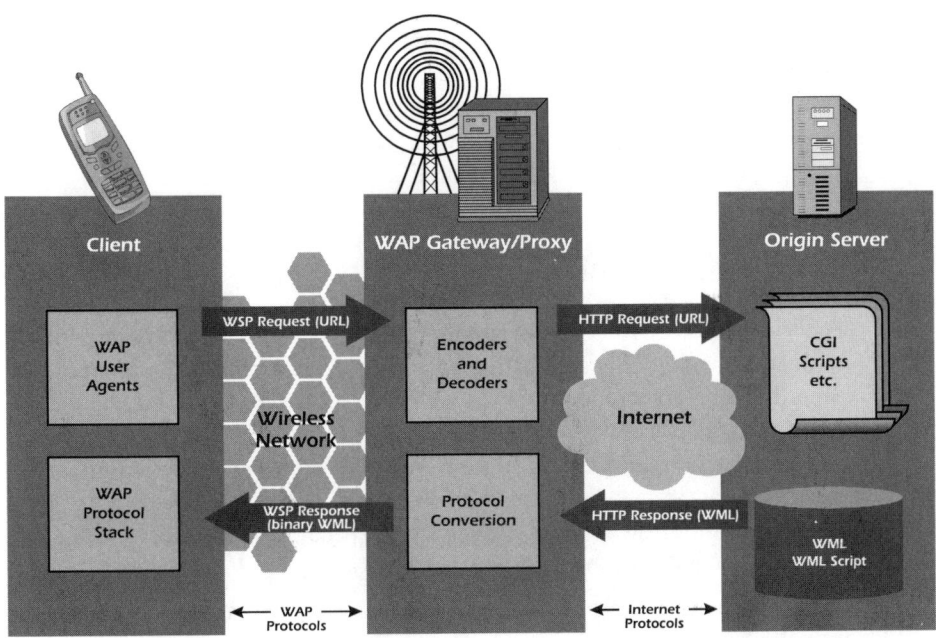

Figure 1.2
WAP architecture and building blocks.

WAP Client

To make use of a WAP service, the user must have a mobile device that is equipped for WAP. Such a phone is equipped with a WAP microbrowser, which fulfills navigation and presentation functions just like Internet browsers on a PC. The navigation function enables the user to request information by entering an Internet address. The browser receives information from the Internet address and presents this to the mobile user. In addition, the browser is equipped with functionality that makes integration of telephone services possible.

Just as there are various suppliers of Internet browsers, there are also several suppliers of WAP browsers. At present, the biggest browser suppliers are Openwave, Nokia, and Ericsson. Nokia and Ericsson developed the browser mainly for their own equipment. Openwave supplies licenses to Motorola, Alcatel, and Mitsubishi, among others. Microsoft also supplies a WAP browser, Microsoft Mobile Explorer, that is used in Sony handsets, among others. Each type of browser presents the information in its own way, depending on the screen size of the apparatus concerned, but not every type of browser supports the same set of facilities. Most suppliers offer a toolkit enabling information suppliers to produce services and then to see how the service will appear on equipment using the browser concerned. Unfortunately, additional testing on the specific equipment concerned is often also required, because the toolkit may not work in exactly the same way as the browser on the cellular phone. The large diversity between browsers can mean that, just as on the Internet, an information supplier must present an application in a number of different ways.

WAP Gateway and Server

The WAP gateway routes information requests from the WAP browser or client to an application server. This application server can be reached via the Internet, but can also be in the same domain as the gateway. The latter is often the situation when a company acquires its own gateway. This has a direct connection with the company's application server (for example, the Intranet server). The information from the application server (the answer to the request) will be decoded by the gateway and sent to the browser. Coding and decoding take place in order to reduce the number of packages that must be sent over the mobile network.

WAP makes use of the same address model as that used for the Internet: Uniform Resource Locators (URLs). The information sought via the WAP browser can thus be found on a server via familiar protocols.

Another possible function of a WAP gateway is a proxy function. Information from the World Wide Web is regularly collected and stored in the WAP gateway server. When the mobile user requests this information, it can be sent directly to the user from the WAP gateway, without collecting it from the Internet. A proxy function is only useful for sites that do not change frequently. Current information can be better obtained directly from the Internet. Some gateways can also translate (x)HTML into WML (Wireless Markup Language, to be further explained in this chapter) and the other way around. This can be useful for information providers who then only need to make their information available in HTML. Some gateways can also be used for NTT DoCoMo's i-mode services offering cHTML in addition to WML. The gateway can be connected to available databases with supporting services.

Supporting Services

The first commercial launches resulted in several additional requirements. Mobile operators needed additional development to offer their customers WAP services. At KPN Mobile, we decided to build an automated provisioning tool using SMS to download initial settings for every type of phone. This was required to lower startup barriers for users without burdening customer care and shops. Also, security and personal profiling requirements resulted in additional investments and undesired propriety solutions.

WAP 2.0 incorporates supporting services, including PKI portals, personal profile standards (UAProf), and provisioning facilities. PKI portals enable end-to-end secure applications, because handsets can initiate creation of public key certificates. UAProf supports retrieval of personal profiles by applications to personalize services to a high degree, without the need for a user profile for every individual application. The provision facilities provide operators a nimble facility for a range of WAP phones instead of a facility dedicated to one type of phone or vendor.

Networks and Protocols

WAP is a protocol that is bearer-network independent. That means that WAP can be used with all common network standards like GSM, CDMA, and PDC. WAP also works with both SMS (Short Message Service) and with a GSM data connection, but it's also suitable for (future) bearers, such as GPRS (General Packet Radio Services) and 3G. The most relevant issue regarding the different networks is the use of circuit-switched networks or packet-switched networks. Packet switching (offered by GPRS and the Japanese i-mode bearer, PDC-P) improves the user experience. The features of the current and future bearers will be covered later in this chapter.

SMS: SHORT MESSAGE SERVICE

> The cellular telephone offers various data communication options. In addition to "normal" data transfer with a speed of 9.6 kbits, use can also be made of Short Message Service (SMS). Text messages with a maximum length of 160 characters can be sent and received via a cellular phone. Every GSM phone is suitable for sending and receiving SMS messages. A help desk worker can also send SMS messages via a PC with special software or via the Internet. Requesting information about traffic jams, share prices, or the weather is no longer a novelty—this has been possible since 1996 using SMS. Company-specific SMS applications have been built to facilitate communication between the office and staff in, for example, the transportation sector or for companies with traveling mechanics.
>
> In spite of these options, the breakthrough of SMS did not take place in Western Europe until the end of 1999. A number of reasons lie behind this late breakthrough:

- Mobile operators have given priority to simplifying the mobile telephony proposition, with the aim of increasing sales. In marketing communication, SMS has remained secondary to speech.
- Until the end of 1999, sending SMS messages with prepaid phones was not possible. This meant that many, often younger consumers could not use SMS.
- The first cellular users saw an envelope on their screen—the sign for a new SMS message on most phones—as an icon for new voice mail messages.
- The telephone number of the SMS exchange had to be preprogrammed in the phone before an SMS message could be sent.
- The cost of sending SMS messages is relatively high.

Many users find composing SMS messages on their cellular phones too complex. Chatboards and predictive text input can make this simpler. Remembering a keyword, inputting, and sending to a specific number to receive news or a file message 160 characters long is asking a lot of the user.

Because cellular operators now aggressively promote SMS and it is also within the reach of a young user group, there has been enormous growth in SMS traffic, from 2 billion messages per month worldwide at the end of 1999 to a total of 14 billion messages per month at the end of 2000, and a steady monthly growth rate of 10 to 15 percent. Nokia even predicts 100 billion messages in December 2003. During a popular German TV show, *Jede Sekunde zahlt* (*Every Second Counts*), viewers were asked to react to questions by sending an SMS message. The SMS system, specially designed by SMS vendor and market leader CMG, handled more than 1.2 million messages in 30 minutes. The use of SMS in Japan quadrupled following the introduction of package-switched networks similar to GPRS. This is a signal that SMS traffic will continue to grow after introduction of GPRS networks to the rest of the world.

The success of SMS has led to the development of EMS (Enhanced Mobile Messaging) and, later on, MMS (Multimedia Mobile Messaging). New functionality like audio and video clips, photographs, and images are added to SMS. MMS is expected to become a mass market service for mobile operators.

WAP

Communication between the WAP gateway and the server takes place via the Internet and the usual protocols. For communication over mobile networks, Wireless Session Protocol (WSP), a binary version of HTTP, is used. To guarantee expandability as well as scalability and flexibility, WAP is set up in five layers. This layered architecture offers other services the possibility of making use of the WAP features, without these services having to be specified in the WAP Forum. The several layers are:

- Application layer, or Wireless Application Environment
- Session layer, or Wireless Session Protocol
- Transaction layer, or Wireless Transaction Protocol
- Security layer, or Wireless Transport Layer Security
- Transport layer, or Wireless Datagram Protocol

In Figure 1.3, the WAP protocol stack is shown in relation to the Internet protocols.

A detailed description of the WAP protocols is outside the scope of this book. The interested reader can consult the WAP specifications at *www.wapforum.org*. We

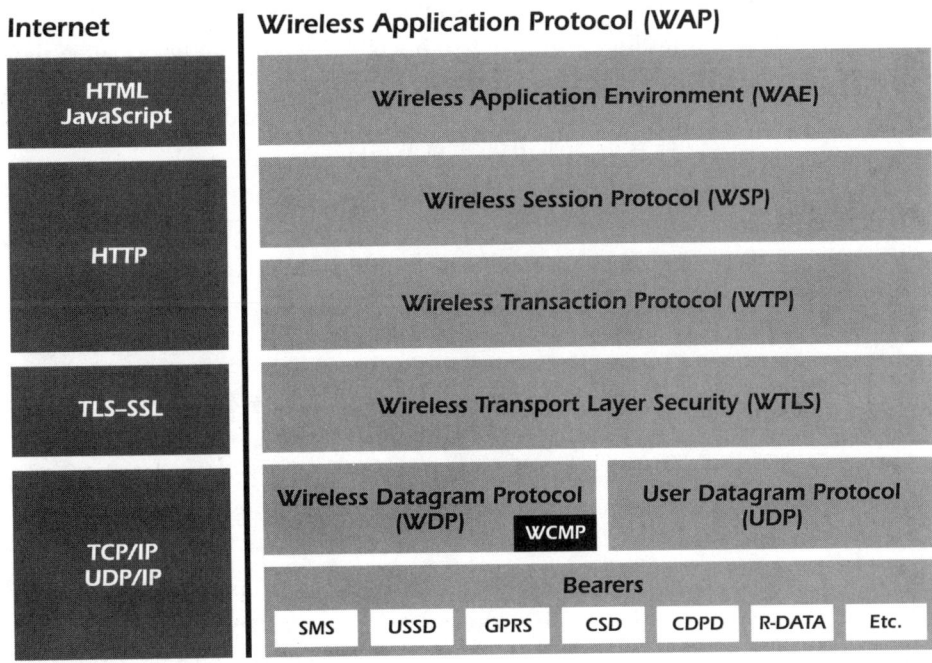

Figure 1.3
WAP protocol stack in relation to the Internet protocol stack.

will go deeper into the application layer, as this forms the foundation for building WAP services and in particular the layout language, the scripting, and the telephony application. The security layer will also be detailed because of the importance of security for payment functionality, for example.

Wireless Markup Language

Services made up in Hypertext Markup Language (HTML)—the language that is used for the layout of Internet pages on the World Wide Web—do not take sufficient account of specific circumstances that arise in a mobile environment. That is why WML was developed. WML is the mobile variation of HTML. A WML file is called a *deck*. This deck is made up of cards. A *card* prescribes how the interaction with the user will appear. Only one card at a time can be displayed on the mobile telephone screen. The following interactions are possible with WAP 1.1:

- *Text.* Text is displayed.
- *Images.* An image is displayed.
- *Flash screen.* With this option a screen is displayed to the user for a period of 1 to 5 seconds, after which the following screen (new card) is displayed.
- *Links.* Just like with standard Internet, one or more links can be shown on a screen. Clicking a link causes the display of another card. This card can be from the same deck. The card comes from a different WML deck if the user clicks on a link to another WML site.
- *Selection.* A selection can be made from a series of options.
- *Input.* The user can be requested to input data. This might be only numbers, only text, or a combination of text and numbers.
- *Soft buttons.* The developer can program functions to run from soft buttons. Every modern device has at least one soft button or menu button.

Regarding layout, there are a few possibilities. The text can be made larger or smaller, bold, or italic using tags. The text can also be left- or right-aligned or centered. Some browsers can display tables. Also, small pictures, called wireless bitmaps, can be one of the options. Examples and tips for the design of your own WAP page are given in Appendix A.

In addition, WML supports soft buttons. These buttons at the bottom of the screen can have their identity changed and you can trigger an action by clicking on them. The left button, Options, is standard and enables options such as Back to homepage and To bookmarks. The right button is often Back, taking the user back to the previous screen.

WML can also handle Events. This is necessary for use with telephone services, for example, if the user has an incoming call.

Considering the limited graphical and interaction possibilities, it is important that the information provider presents the data as concisely as possible. After all, reading long pieces of text on a mobile telephone is not easy. Also, the amount of input from the user must be limited as much as possible, because typing data with the keyboard of a cellular telephone is also not easy. Suppliers of WAP telephones are doing their best to make surfing as simple as possible. On the Nokia 7110, there is a NaviRoller™ to enable easy scrolling through text and easy clicking on links (see Figure 1.4). Ericsson has a touch screen on the R380. Links can be easily clicked using a pen.

Figure 1.4
The NaviRoller™ of the Nokia 7110.

Japanese Competition for WAP?

> The Japanese market for mobile Internet is the biggest and most advanced in the world. Since April 1999, Japanese mobile operators have offered mobile access to Internet services. NTT DoCoMo is easily the market leader, with more than 7.5 million mobile Internet users by the end of June 2000 and up to 19 million users by the end of January 2001. More than 90 percent of new subscribers start using i-mode (see Figure 1.5). Competitors IDO and J-phone are quite successful marketing WAP services, too. They have been offering WAP services with platforms delivered by U.S. company Openwave. IDO's WAP service, named EZweb, was launched in April 1999. After a slow start, the subscriber base grew to more than 3 million by mid-2000, adding 1 million subscribers during the last 2 months. At that moment, one in three IDO customers used their mobile phones for WAP services.

NTT DoCoMo uses Compact HTML (cHTML), a subset of HTML, rather than WML for its i-mode service. This makes WML-based converters unnecessary and it is thus even easier to build mobile Internet sites. Midway through 2000, 15,000 Web sites had been developed in cHTML and at the end of January 2001, 38,000 sites were available. All these sites are especially for NTT DoCoMo customers. At the same time there were about 10,000 WAP sites from more than 95 different countries. In June 2000, Logica was the first to introduce a gateway (the m-WorldGate) that supports cHTML as a markup language in addition to WML. Also, different browsers are needed and thus a separate mobile device.

cHTML devices in Japan support, among other things, possibilities not supported by the current WAP phones, such as color pictures, GIF animations, and the downloading of ring tones of MIDI quality (see Figure 1.6). Handsets equipped with cHTML will become available in the European market. These will be suitable for both GSM and GPRS. It could be that cHTML will eventually oust WML as the primary markup language for mobile Internet telephones. Since early 2000, in view of the development of WAP services in Europe and the acceptance of WAP as a standard, it could also be that the opportunities for cHTML as an alternative for WAP lie mainly outside Europe. Coexistence is the other alternative. NTT DoCoMo has already announced it will support WAP 2.0. standards as well. DoCoMo also has announced that they would like to export i-mode.

Figure 1.5
Email via i-mode.

Another difference between WAP and NTT DoCoMo's i-mode in Japan is that instead of a GSM network as carrier, PDC-P is used. PDC-P networks have the same data speed as GSM networks, but already offer billing for the amount of sent data instead of connection time and "always-on" functionality. More details about networks and their features are described in the next paragraph. In mid-2000, DoCoMo implement-

Figure 1.6
Japanese phones.

> ed Java and Japan is the first country in the world launching 3G services in 2001. Japan is approximately two years ahead of Europe in data speed. Success will be achieved thanks to the content available. The content available in Japan is only partly usable in Europe. DoCoMo, with its service concept and experience with millions of mobile Internet users, has a know-how advantage over Europe. After all, the Japanese succeeded in marketing this brand new concept very well. NTT DoCoMo bundled new phones and a broad range of easy-to-use services for acceptable prices. KPN Mobile, as a DoCoMo partner, will make use of this. There is more information about the services offered by i-mode in Chapter 5.

WML Script

WML Script makes it possible to improve services that have been written in WML. It can add specific intelligence to a service, such as calculating logical and conditional functions. WML Script can, for example, be used for validating the user's input. Without the use of WML Script, validation must be done in the server and that means a question-and-answer session over the network. With WML Script, a local function can also be used in the mobile device, for example, a telephone function.

Telephony Services within WAP

Wireless Telephony Application (WTA) is also defined in the WAP standard (see Figure 1.7). WTA is an environment that makes it possible to use telephone services. The WTAI (Wireless Telephony Application Interface) offers a collection of telephony-related functions in a mobile telephone that can be activated with WML or WML Script. These are functions related to a call, such as making, breaking, or putting a call on hold. Functions for handling text messages or for the control of the telephone directory in the device can also be activated using WTAI.

Because the real-time behavior of a service is very important for telephone services, it is possible to save specific WTA services in the device. This enables the device to react directly to "events" because it does not first have to refer to a server via the network. Typical events are incoming calls, breaking off a call, and answering a call. The WTA services that are stored in the device can react to these events immediately. The user can receive an indication about the event, (for example, a new voice mail message), and is given the opportunity to start an accompanying service. A user who receives a voice mail indication can choose to listen to this message immediately or save it for later.

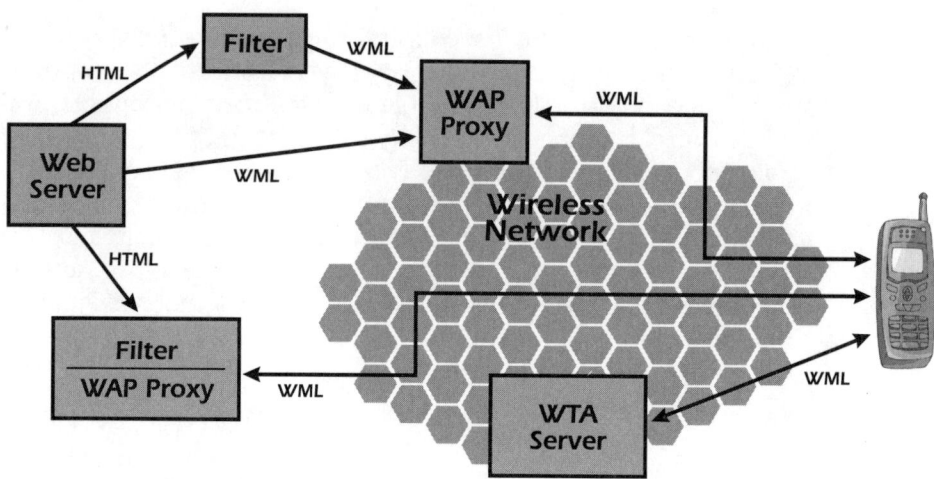

Figure 1.7
The WAP infrastructure with WTA server.

Content Billing

During the emergence of the Internet virtually all information was free. Attempts have been made with varying success to obtain payment for information, by means of a subscription, for example. In many cases access to the restricted Internet site is part of a bigger picture. In this way, subscribers to a magazine can receive access to the archives of the magazine via the Internet plus some additional services. Bank customers sometimes have to pay to access their account via the Internet. The WAP standard has no facilities for billing customers via the mobile bill or the prepaid account, so operators develop their own facilities, depending on the business model they choose to offer. A few mobile operators offer information suppliers payment facilities if the information is supplied via WAP. @info from KPN Mobile gives information suppliers the facility to ask a maximum of around $1 per day (or per successful information request) to be included on the mobile invoice. NTT DoCoMo has also developed billing services for 1,200 "recognized" sites to charge end users via their mobile phone bill for using the service.

Other mobile operators buy information for a fixed sum and charge for this information within the tariff for the GSM connection. A fixed tariff is charged irrespective of the value of the information. A third possibility is that the information is sponsored by advertising income.

MOBILE INSTANT MESSAGING

In a mobile environment, messaging applications are very important. The success that SMS and email notification services already had before WAP was introduced proved this. The WAP 2.0 framework facilitates building messaging applications in different content formats, including images, audio, video, animation, and all kinds of data records.

Another development in this area is the instant messaging standards war that is being fought at the moment against America Online (AOL). Instant messaging makes it possible to see which of your friends are online, so that you can send them a short message, sound, or picture. This is an excellent example of a WTA "killer app." With ICQ (I seek you) and Instant Messenger (over 20 million active users), AOL controls this market of more than 1 billion messages per day. Yahoo! and Microsoft's instant messaging programs each have more than 10 million active users using the fixed Internet. The different programs are not interoperable, so most heavy users use more than one program. Tegic is the inventor of T9 software, which makes it possible to send simpler messages from a mobile telephone. Tegic—acquired by AOL—has developed a mobile ICQ. Customers can ICQ with a friend regardless of whether they are using a fixed or mobile Internet device. On top of this, AOL has integrated SMS in its ICQ window. The WAP 2.0 framework includes facilities to support multimedia messaging, including email and instant messaging programs on WAP handsets.

Security

Security is very important with all types of services that include banking and payment, regardless of the medium used. The following aspects are important:

- *Identification.* Who is the user of the service? How can the user be recognized?
- *Authentication.* Are the users really who they say they are? The same is of course equally valid for the service supplier.
- *Authorization.* Is the user authorized to make use of this service?
- *Confidentiality.* Is the information being exchanged secure from being read by unauthorized persons?
- *Integrity.* Is the information secure from alteration during exchange?

- *Irrefutability.* Is it impossible for the participating parties to negate the transaction or part of it?

In the WAP Forum, Wireless Transport Layer Security (WTLS) has been developed. WTLS looks after the following security components:

- *Encryption.* The transmitted data is encoded between the browser and the WAP gateway, so that the confidentiality of the information transfer can be guaranteed.
- *Data integrity.* The data that is received by the browser or the gateway is controlled by control bits to ensure that it is the same as that sent by the gateway to the browser.
- *Authentication.* Authenticity of server and client is guaranteed by digital certification. In the WAP 1.1 standard, only server authenticity is provided. WAP 1.2 extends this to the clients.

This functionality gives the assurance that the information sent has not been manipulated by a third party, that privacy is safeguarded, that the author of a message can be identified, and that both parties cannot deny that they have exchanged information.

Nevertheless, this is still not sufficient for high-security applications. WTLS only looks after security between the browser and the WAP gateway. This means that there is no end-to-end security between the client (the browser on the mobile telephone) and the application. After all, the security between the WAP gateway and the application server is not defined by WTLS. The connection between the WAP gateway and the application server can be the Internet, but use can also be made of HTTP-S (an encoded channel on the Internet), ISDN, or a fixed line. Safe applications can be made using the latter connection, where, thanks to the security of the GSM network, a higher security level can often be reached than with the analog dialing of a server.

The WAP Forum has accepted proposals to take up the Wireless Identity Module (WIM) in the WAP standard (WAP 1.2) to achieve end-to-end security. WIM is used for performing WTLS and application-level security functions. Information concerning user identification and authorization often requires sensitive data (keys). The keys and the operations involving these keys can be stored and handled in the WIM. To realize the required level of security (tamper resistance), physical hardware protection is used. Mobile phones and PDAs do not reach this security level. Smartcards (e.g., SIM cards placed in every GSM phone) are suitable to perform the WIM functionality. The SIM (Subscriber Identity Module) is the smart card that must be placed in a GSM device to enable the user to access GSM services. The SIM ensures the unique identity of the GSM network user. WIM enables extra security due to RSA signing, private key decryption, and storage of certificates for user, server, and applications.

An application uses WIM for signing and unwrapping of a key. The private key will never leave the WIM and the operations are generic, so any application can make use of the facilities provided by WIM. Signing may be used for authentication, signing of documents, and confirmation of transactions. The user may be asked to enter a Personal Identification Number (PIN) for every signature made. Also, the WIM can be used by the application to calculate a digital signature using the private key.

If an application receives a wrapped key, it will send it to the WIM. WIM deciphers it using the private key and returns the unwrapped key. The application may use the unwrapped key to decode the attached message.

Existence of WIM on the SIM card has no effect on the GSM functionality. If the mobile phone does not support WIM, the SIM can be used for GSM purposes (e.g., making and receiving calls and WAP services not requiring WIM functionality). WIM will be added to the SIM during the personalization process and by using Over The Air mechanisms. With WIM as a basis, Nokia, Ericsson, and Motorola have once again joined forces in a joint venture called Mobile Electronic Transactions (MET), which will develop an open standard for mobile payments.

Does this mean that safe applications are not possible without WIM? That is not the case. Secure connections of a GSM device to a WAP gateway with WTLS and from the WAP gateway to the application server with HTTP-S are already possible and if the mobile operator becomes familiar with them, various options for payment are already available. Online bookseller Bruna and online CD vendor Boxman already offer the option to purchase books and CDs via credit card payment using WAP.

SIM Application Toolkit

The SIM application toolkit (STK) has been developed to make it easier to send SMS messages. Menu structures can be added to the menu of the mobile telephone using a new type of SIM card. The advantage of this is that the user does not have to type in or remember a password. Another advantage of services on the SIM card is that this technique is adequately secure for bank transactions. STK supports digital signatures from the SIM card. The disadvantages of SMS remain: the limited message size of 160 characters and the delays that can occur during sending and receiving messages. For each transaction, several messages must often be sent backward and forward and this can take considerable time and lead to high costs. The fact is that every message sent costs money. The programming of menus on existing SIM cards makes extension or changing of the menu difficult. The SIM card will have to be replaced often. New JAVA-card 2.1 SIMs promise interoperability between different vendors and an ETSI-approved download standard. This means that applications can be downloaded to the SIMs through the network without physical return or replacement of the SIM. New class 3 functionality offers the SIM control over calls and sending of SMS messages.

The SIM is important for a mobile operator because the operator issues the card and determines the applications on the card. Handsets are often packaged to a SIM by the outlet. Therefore, the operator does not control type and software release of the handsets. The improved functionality and flexibility of the SIM card opens up new opportunities for operators.

In the Netherlands, most operators use the STK to stimulate use of SMS messages by offering a customer menus instead. Dutchtone has introduced I mobile. This is a SIM toolkit application enabling popular services like news, traffic, and stock prices to be accessed easily. BT Mobile's Telfort has also launched a SIM toolkit application that enables email messages to be read. Both applications have been built to improve the user interface for SMS information services and not specifically for high-security applications. In the United Kingdom, more than 100,000 customers of Barclays Bank make use of a SIM toolkit banking application.

WAP via a GSM data link overcomes the 160-character restriction and supports graphics. The dynamic menu structure makes expansion and alteration of services possible as well as online interaction without delays. This makes use of services both simple and direct. Because of the limited penetration of WAP telephones compared to the 100 percent penetration of SMS, the offering of SMS services could be very attractive over the next two years. In many instances, current SMS users will be the WAP users of the future. Also, the experience with SMS services gained within the organization is applicable to WAP. With SMS, the need for concise messages is even greater and topicality and speed are preconditions.

Companies with higher security needs such as banks, can make use of the STK. STK makes it possible to place extra applications on the SIM card of a mobile telephone. These can be applications to simplify the operation of services such as voice mail and SMS information services, or programs for mobile banking or electronic signatures. These applications make use of SMS as a carrier. A suitable device is needed for use of SIM Toolkit. All current devices, including WAP devices, support STK, but many people have older devices that are not suitable.

In principle, WAP services could start offering the customer all sorts of options via WAP, but with payment via SIM toolkit. This depends on what operators put on their SIM cards. At the moment, most SIM cards have very limited memory (8, 16, 32, or 64 kbit) and therefore cannot support extensive programs. The operator has to choose which applications are important enough to be placed on the SIM card. Subsequently, for good market penetration of these new SIM cards, it will be necessary that many of the existing cards be replaced, leading to higher costs for the mobile operator. It is thus probable that many operators for WAP will wait until WIM is specified and place this on their cards for new customers. Until then, they will cooperate with banks and payment services for their existing customers on the basis of SMS.

WAP and USSD

USSD (Unstructured Supplementary Services Data) is another carrier that can be used for sending messages with a maximum length of 182 characters. USSD, just like SMS, uses the signaling channel of the GSM network as a carrier. With SMS, every time there is a message, a channel is reserved. With USSD, a session only starts at the moment that the user starts to use a USSD service and the radio connection remains open until the application user finishes the session or until a time-out takes place (if there is no data transmission for a specific period). This makes USSD seven times faster than SMS during a session. Nokia is offering a platform where USSD can be used as a carrier for information services complementary to the SMS services. This can be interesting, particularly for services that need fast response times, such as chatting. USSD can also be used as a carrier for simple WAP services next to it.

WAP on the SIM Card

WAP on the SIM makes the purchase of a mobile telephone with a WAP browser unnecessary. WAP services can be used via existing devices with a new SIM. The advantage of this solution is that the scarcity of WAP telephones will not prevent the introduction of WAP services. Also, it is not necessary to replace good devices that do not have a WAP browser. Another advantage of services on the SIM is that this technique is adequately secure for bank transactions. STK supports digital signatures from the SIM, enabling end-to-end security to be offered.

WAP on the SIM requires a device that supports the STK. The limited memory of the SIM has negative consequences for WAP services. The use of graphics is not possible. The menus cannot be extended and are limited in number and depth (the number of menu layers). There are also no soft key options or options for building in intelligence that would make personalization possible.

Implementation of a limited set of WAP functions on the SIM is not interesting for the innovator. The innovator will not accept a product with restrictions. The remaining interested users become confused—WAP, SIM toolkit, WAP on SIM card, GPRS, and so on. These users are interested in services based on techniques that will not become outdated within a year. The implementation of WAP on the SIM has consequences for the mobile operator. The distribution of SIMs to users is neither easy nor cheap. Besides, the operator must invest in an infrastructure specifically for WAP on the SIM.

Considering that the shortage of devices is temporary, end-to-end security in the device for WAP browsers is close, and the service for users, suppliers, and operators has many drawbacks, it seems that this is a short-term solution for mobile operators. Virgin is a virtual mobile operator because they make use of the mobile network from

One2One. Virgin does supply SIM cards themselves. As the fifth operator in a market nearing saturation, a large number of Virgin customers will quit their service with another operator and keep their device. Virgin has offered WAP on the SIM to customers who wish to use WAP but do not want to purchase a new device. WAP on the SIM does give parties such as Virgin the possibility of enabling users to call on their own WAP portal without the device having to be financed.

CELLULAR NETWORK STANDARDS

WAP is designed to work with most known mobile network standards for mobile telephony, two-way paging, and two-way radio. At first this seems rather trivial but it is not. There's a variety in standards used by different countries. WAP can be used on the following networks: GSM, CDMA, CDPD, TDMA, PDC, PHS, Flex, ReFlex, iDEN, TETRA, DECT, DataTAC, and Mobitex. It's beyond the scope of this book to describe each standard in detail. We will focus on the main standards for mobile telephony, the main differences, and their upgrades to facilitate mobile Internet services now and in future networks, often referred to as 3G (third generation).

The other chapters of this book often refer to GSM, GPRS, and UMTS because these are the European standards. With respect to mobile Internet services this is not relevant because of the bearer-independent nature of WAP and its successors.

First-Generation Mobile Networks

One of the first analog standards is called NMT (Nordic Mobile Telephony). NMT was introduced in 1969 by the four Nordic countries and provided roaming possibilities (using the same mobile phone in a different network). NMT evolved and was rolled out in Switzerland and the Netherlands as well. Other well-known analog standards are Advanced Mobile Phone Service (AMPS) and Total Access Communications System (TACS) standards. In general, these standards are named first-generation systems (1G).

Mobile telephony standards are about technologies using available radio spectrum. Different types of cellular systems employ various methods of multiple access. The traditional analog cellular systems, such as AMPS and TACS, use Frequency Division Multiple Access (FDMA). FDMA channels are defined by a range of radio frequencies, usually expressed in a number of kilohertz (kHz), out of the radio spectrum.

For example, AMPS systems use 30-kHz "slices" of spectrum for each channel. Narrowband AMPS (NAMPS) requires only 10 kHz per channel. TACS channels are 25 kHz wide. With FDMA, only one subscriber at a time is assigned to a channel. No other conversations can access this channel until the subscriber's call is finished, or until that original call is handed off to a different channel by the system. AMPS is still in use in the United States. TACS is still in use in European countries like Italy, Spain, and the United Kingdom with a few million customers. NMT is no longer in use in the Netherlands, where the spectrum is used for GSM. In the Nordics and other countries, NMT is still operational but the amount of users is decreasing quickly in favor of the GSM networks. WAP is not supported by these networks.

Second-Generation Mobile Networks

The next-generation networks have been rolled out in Europe since 1991. The dominant 2G standards are GSM, CDMA, TDMA, and PDC. Starting in the Nordic countries, GSM (Global System for Mobile Communication) has become the dominant standard. GSM is available in more than 160 countries and offered by more than 400 operators. With about 500 million users globally, GSM had captured about 70 percent of the wireless market by the first quarter of 2001. GSM is dominant in Europe, the Middle East, and Africa and is growing strong in Asia Pacific. The second most important second-generation standard is CDMA (Code Division Multiple Access). With more than 80 million subscribers, it's dominant in the United States and has a presence in Asia Pacific, the Caribbean, and Latin America as well. Another digital standard, TDMA (Time Division Multiple Access), was used by 61 million subscribers in North America and Latin America by the end of 2000. TDMA claims to be the most widely used wireless technology in the Americas. The Japanese developed their own standard called PDC (Personal Digital Cellular), which is used by NTT DoCoMo and a few other Asian networks.

TDMA is a common multiple access method employed in new digital cellular systems. TDMA digital standards include North American Digital Cellular (also known by its standard number IS-54), GSM, and PDC. TDMA systems commonly start with a slice of spectrum, referred to as one carrier. Each carrier is then divided into several time slots or channels. Only one subscriber at a time is assigned to each channel. No other conversations can access this channel until the subscriber's call is finished, or until that original call is handed off to a different channel by the system. For example, IS-54 systems, designed to coexist with AMPS systems, divide 30 kHz of spectrum into three channels. PDC divides 25-kHz slices of spectrum into three channels. GSM systems create eight time-division channels in 200 kHz-wide carriers. With CDMA, unique digital codes, rather than separate RF frequencies or channels, are used to differentiate subscribers. The codes are shared by both the mobile phone and the base station. All users share the same range of radio spectrum.

Interoperability between the different standards is one of the big issues. Using the same phone and the same SIM card wherever you go becomes common to most Europeans. Using your GSM phone in other networks (or vice versa) is not that easy. Different ways of global roaming exist: from having a "world phone" (or a rental phone while keeping your SIM), which can be used in some American networks, or another mobile phone. Sometimes it's possible to keep your own number and receive the costs on the same bill. In many cases another phone number is required. They all have two things in common: It's a hassle and it's expensive. One of the objectives of 3G is realizing true global roaming. In the meantime, integration takes place to make global roaming on 2G networks easier by providing multiband (phones that can make use of different frequencies) and multisystem (phones that can use more than one mobile standard) phones, SIM cards, and roaming agreements.

WAP and 2G Mobile Networks

Most commercial WAP services make use of a data link as bearer, and that is a circuit-switched connection. For a WAP session, a connection is set up that remains open during the whole session. Thus, the connection remains open even when the user is reading or entering data. This is a disadvantage of the circuit-switched data link as carrier. The other big disadvantage is the 15 to 30 seconds needed to make the connection. In November 1999, KPN Mobile was the first GSM operator offering services based on WAP 1.1 services. In the meantime, most other European operators offer WAP services. In about the same period, operators across North America, Korea, and Japan began launching cdmaOne Internet and information services.

As mentioned before, it is also possible to make use of SMS as a bearer service in the place of the data link. New WAP menus and information are then collected by means of SMS messages and input data is sent via SMS. A disadvantage of this option is that the interaction is certainly not immediate. If the SMS exchange receives a lot of messages to be sent, a queue of SMS messages develops, causing delays and unreliability. The first WAP implementations, for example, the WAP browser in the Siemens S25 telephone, could only make use of SMS as the carrier. This was not successful when it became apparent that most operators made use of circuit-switched data for connection to the WAP gateway.

With a GSM data link, circuit switching is at the disposal of the user during the entire connection time. The bandwidth of 9.6 kilobits per second (kbps) is limited and the time needed to establish the connection is quite long. A few European operators, such as the German E-Plus, have built facilities into their networks to enable use of a maximum of four channels simultaneously for data transmission. This technique is called High Speed Circuit-Switched Data (HSCSD) and makes data speeds of 57.6 kbps possible. Nevertheless, most operators choose GPRS because they do not have the extra capacity in their networks necessary for a successful HSCSD offering. The Nokia Communicator 9210 makes use of HSCSD. According to Nokia, every country has an operator offering HSCSD and high-end users demand proven technology. According to Yankee Group's estimation, Nokia will miss 85 percent of the market.

2.5-Generation Mobile Networks

The market demand for the full complement of 3G services is not yet quantified. Similarly, spectrum allocation differences across regions and continents make a global 3G band seem unlikely. Meanwhile, today's 2G systems are still rolling out. Operators from all technologies have already invested billions of dollars in state-of-the-art infrastructure. Operators appreciate 3G's potential, but their financial reality is in today's investments in today's market. The packet-switched carrier services like PDC-P, CDMA2000, and GPRS overcome the current disadvantages of 2G networks. Because they are extensions of the second generation, they are often called 2.5 generation (2.5G) networks.

In contrast to circuit switching, package switching does not make a connection with a specific capacity that is only for the exclusive use of the user. The information is placed in packages and these are sent over the network. Packages from various senders make use of the same transmission line. For speech and other applications where delay leads to excessive loss of quality, circuit switching with the current GSM speeds is the best solution. Package switching is ideal for data traffic.

Package switching has the following advantages:

- *Efficient use of the bandwidth.* If no information is transmitted for a short period with a circuit-switched connection, the open line remains. With package switching, this bandwidth can be utilized for packages from other users.
- *Always online.* The user does not have to make a connection every time that he or she wants to send or receive information. Once the user has made contact with the server used for sending and receiving data, he or she can remain online all day and only pay for the actual use made.
- *Paying for use.* With circuit-switched connections, the user is charged on the basis of the length of time that he or she has made use of the connection, regardless of the quantity of data transmitted. With package-switched data transmission, it could also be possible to charge on the basis of the number of bytes sent and received.

GPRS

General Packet Radio Services (GPRS) is the standard for package switching in GSM networks. GPRS offers higher speeds, thanks to the more efficient use of bandwidth. The expectation is that when introduced, GPRS terminals will provide a speed of 56 kbps in the direction of the user and 14 kbps from the user to the server. This higher speed, compared to the 9.6 kbps of the current GSM traffic, offers a broadening of the number of types of possible applications and improved user friendliness for the current GSM data applications. In the first instance, GPRS seems to be too slow for real-time video and multimedia, but will enormously stimulate the development and use of WAP services. GPRS is in this respect not a competitor of WAP, but a complementary technology. To be able to offer GPRS, operators must adapt their networks. This means considerable investment. Thus, the British mobile operator Orange paid around $60 million to telecom manufacturer Ericsson to make the Orange network suitable for GPRS. In late Summer 2000, Telstra (Australia) and Smarttone (Hong Kong) introduced GPRS services. During 2001, most European operators rolled out their GPRS networks. As with WAP, GPRS services might not meet customers' expectations. Speeds are often lower than the maximum 56 kbps and the limited time window of GPRS might restrict companies to invest heavily. Most European operators first introduced their GPRS subscriptions to the corporate market for access to corporate intranets and applications because of the limited availability of GPRS terminals in 2000 and the first half of 2001 (see Figure 1.8). Another reason for this is that GPRS is a complex new network technology and most European operators thought it wise to start experimenting with a limited set of heavy-using clients instead of introducing it to consumers when the technology is not fully tested.

Figure 1.8
Examples of GPRS phones by Sendo (Z100) and Trium (Geo GPRS).

GPRS Terminals, Traditional and New

A GPRS terminal can be a mobile telephone, possibly with a WAP browser. A GPRS terminal can also be a modem card that fits in the PCMCIA slot of a laptop. Handhelds and palmtops will probably make use of WAP for access to the Internet. Laptops will remain using traditional HTML Web browsers and their successors for Web browsing. All types of Web browsing will become much more user friendly with GPRS because of the higher speed and the fact that it will no longer be necessary to dial in. Also, new types of GPRS terminals will come onto the market, enabling, for example, drawings, photos, and moving pictures to be made and exchanged or high-quality sound files to be made and subsequently transmitted.

Standard GSM telephones are not suitable for GPRS. Also, mobile telephones with WAP browsers are unfortunately not suitable for GPRS. Customers have to buy a new phone. Delay in the introduction of new handsets forced many operators to stick to commercial pilots longer than they intended to.

PDC-P

PDC-P is a package-switched carrier service developed by the Japanese. It is comparable to GPRS. PDC-P has a minimal dial-in time and the service is always on standby after the customer has dialed in. Use is calculated per amount of data sent and not per minute. The data speed at 9.6 kbps is lower than GPRS and is the same as the European GSM network without GPRS. PDC-P is used by NTT DoCoMo for its successful i-mode service.

CDMA2000X

The CDMA2000X proposal (also known as Wideband CDMAOne) is merely an extension of CDMAOne. CDMAOne evolution promises operators progress toward high-speed data in manageable steps: 14.4 kbps is available in CDMAOne already, as well as 64 kbps. Data rates beyond these are already in trial and are being demonstrated by infrastructure providers. Because CDMA2000 evolution builds on the same technology framework, operators have the flexibility to upgrade in a cost-effective manner.

Next in the evolution path is an upgrade that is in the definition phase at this time within the TIA standards process. Implementing this next phase of CDMAOne, an operator can offer services of 144 kbps on a 1.25 MHz channel on its current system. This data rate exceeds what's available to most worldwide Internet consumers today.

Third-Generation Mobile Networks

Third-generation networks promise the most appealing services, like surfing the Web quickly with mobile phones and laptops, viewing electronic maps, and transferring files from PCs.

3G covers the standards EDGE, UMTS (or Wideband CDMA), and CDMA2000 3x. CDMA2000 3x is an extension of the CDMA2000 technology described in the previous section. CDMA2000 3x and EDGE both use already allocated spectra. Auctions for the third-generation mobile spectrum have been taking place in Europe and will be held during 2002 in the United States. The U.S. Federal Communications Commission adopted a Notice of Proposed Rulemaking to explore the possible use of frequency bands below 3 GHz for 3G. The most coveted spectrum is now used by the U.S. Defense Department for satellite tracking and radio communication. Sharing spectrum with 3G services would cause interference and moving defense services to other frequencies would be too disruptive. The mobile industry focused on global consistency of frequency bands, meaning using the same mobile device anywhere. If this cannot be realized, rollouts should be delayed and handset and infrastructure prices will increase. Handsets with networks on various frequencies already exist and can be improved, so a scarcity of frequency in the United States seems to be a bigger potential problem if 3G services prove to be successful.

EDGE

EDGE (Enhanced Data Rates for GSM Evolution) is a technique that makes it possible to send at 64 kbps instead of the present 9.6 kbps over a single GSM or TDMA channel. The expectation is that this technique will be implemented in mobile networks during 2002. The maximum performance for speech and data services can be extracted from the network when it is used in combination with the packet switching of GPRS. EDGE creates a migration path for GSM from GPRS to UMTS (Universal Mobile Telecommunication Services), because the changes to be made will later also be useful for UMTS. The success of EDGE will be determined by the timely availability of mobile devices (see Figure 1.9) and by the introduction of UMTS. The longer the UMTS introduction is delayed, the greater the chance that operators will implement EDGE. EDGE can perhaps increase the survival chances of operators that fall by the wayside during the auction of UMTS licenses. EDGE is standardized within the ETSI (European Institute for the Standardization of Telecommunication) for GSM networks and was later taken on by ANSI for CDMA networks in the United States.

The introduction of GPRS and spectrum auctions have resulted in major shifts in North America and Latin America. The U.S. auctions opened up new spectrum opportunities to GSM and TDMA players. AT&T Wireless announced it was deploying GSM/

Cellular Network Standards

Figure 1.9
Prototype of an EDGE phone by Ericsson.

GPRS alongside its existing TDMA network. AT&T previously stuck to a migration from TDMA to EDGE, but the decision to introduce GSM/GPRS enables the operator to start large-scale wireless data services by the end of 2001. GSM's clear path to 3G via GPRS is an important reason for other TDMA operators to start offering GSM services. Driven by their European shareholders, El Salvador's CTE Personal will replace its TDMA network for GSM and Bolivia's Entel Movil will build a GSM/GPRS network alongside its TDMA network, and Telcel Mexico will do the same. The development of TDMA via EDGE is uncertain compared to GSM/GPRS. Credit First Boston announced it may signal the death knell for EDGE, with a diminishing market reducing terminal demand.

UMTS/W-CDMA

The next step in the mobile evolution (see Figure 1.10) is UMTS, which will use new radio technology with speeds of between 384 kbps and 2 Mbps. The radio technique that will be applied for this in Europe is Wideband Code Division Multiple Access (W-CDMA). With the high speeds offered by UMTS it will be possible to send moving pictures and a large number of mobile applications will be extensively expanded with that option. In addition to the new radio part, the UMTS network makes use of the same circuit- and package-switched exchanges as the GSM network. Mobile operators who manage to acquire a UMTS license will be able to use the investments they have made in GSM and GPRS toward UMTS. Acquiring licenses and building networks will cost the European mobile operators between $275 billion and $350 billion. NTT

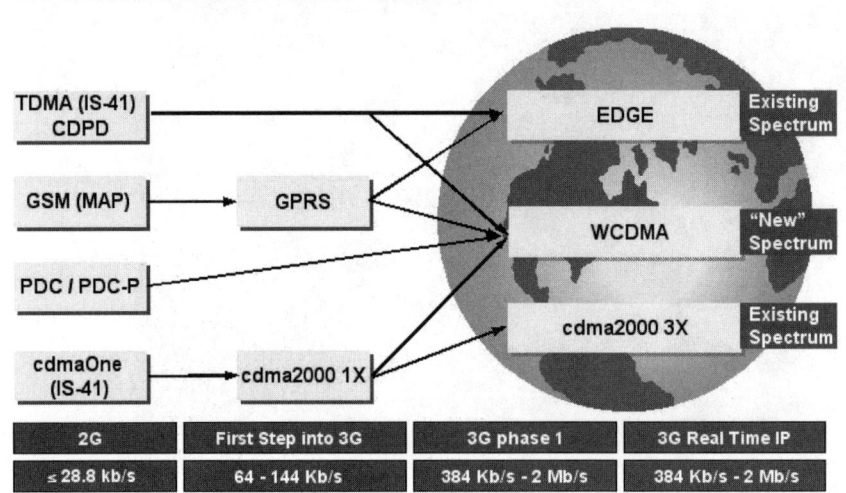

Evolution to 3G / IMT-2000

Figure 1.10
Evolution from second- to third-generation networks.

DoCoMo was the first mobile operator to introduce 3G during 2001. Its competitors will follow in 2002, offering NTT DoCoMo the possibility to expand the lead it already has in mobile Internet. Japanese experience with 3G standard W-CDMA has proven that the limits of available radio spectrum mean that the technology cannot simultaneously provide data speeds needed for long video and sound transmissions to a large number of users. Downloading MP3 files takes a long time and might become quite expensive (see Figure 1.11). This is especially the case in densely populated areas. UMTS offers new mobile capabilities like the following (see Figure 1.12):

- Voice quality comparable to wireline
- Multimedia high-speed data connections
- User customization, like quality control (treble/bass), enhanced control on personal voice mail routing, announcements, call forwarding, call screening, and so forth
- Personalized mobility services; location-dependent advertising and information services, emergency services, and vehicle navigation
- Wireless PSTN and wireless public data network access

Cellular Network Standards

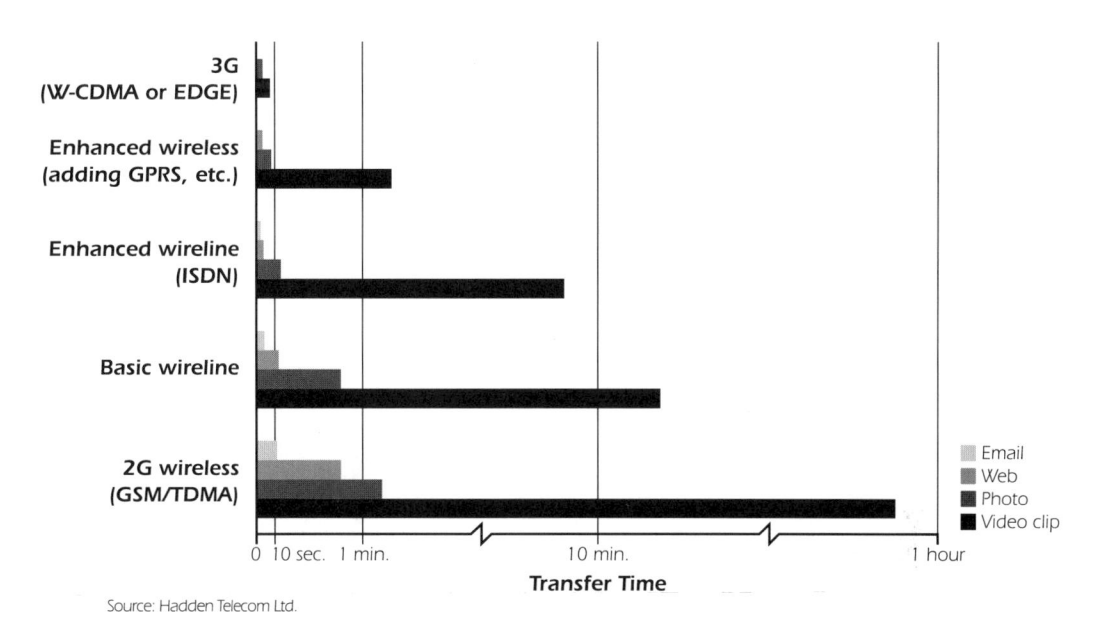

Figure 1.11
Transfer time needed to load different applications for UMTS, GPRS, ISDN, and GSM.

The high investments and the success of the business models NTT DoCoMo introduced with i-mode (see Figure 1.13) will drive operators to open their network for other parties. The expected growth in voice traffic will not cover the expenses, so data will be a key revenue driver. Other parties will offer a broad range of applications combining voice and data over a mobile phone or laptop.

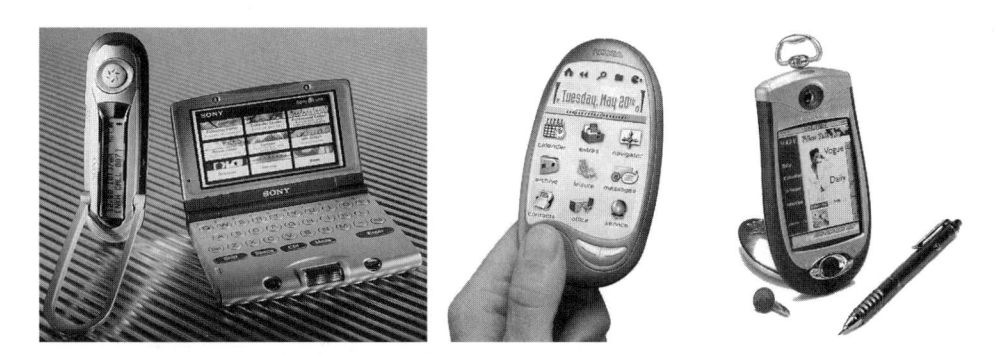

Figure 1.12
Prototypes of UMTS terminals by Sony, Nokia, and Ericsson.

Figure 1.13
NTT DoCoMo advertisement featuring Keiji Tachikawa, President and CEO of NTT DoCoMo.

ERICSSON: THE POWER OF MOBILITY

(By Patrick Blankers, specialist in mobile data communication at Ericsson Telecommunication)

The Internet as it is now is a fantastic phenomenon. It provides access to a worldwide treasure of information, an inestimable quantity of online services, and a virtual shopping mall for the global economy. The Internet is at the threshold of a new phase: mobile Internet.

In contrast to the fixed Internet, mobile Internet can always accompany us. The personal and mobile character of Internet services will certainly change our lives. It will bring us better ways of getting in contact with our friends, it will be a "wireless purse," and it will make personal information accessible—irrespective of time and place. In the near future, interactive games, music, movie clips, and other forms of entertainment will be within reach via a pocket terminal.

Positioning is one of the most valuable characteristics of mobile data services. The GSM network is capable of delivering data that can determine the user's position. Recently developed techniques make positioning increasingly more accurate by determining the distance of the user from three local GSM transmission masts. If the position of the Internet user is known, a large number of tailor-made information services can be offered. "Where is there a hamburger restaurant in the area?", "When does the next bus leave?", and "How old is the building that I am now standing in front of?" are all questions that can be answered instantly via mobile Internet access. WAP has been specially developed for such information services. The examples just given assume that the initiative lies with the GSM user, but that need not always be the case. With so-called push services, it is the service provider who sends information to the GSM user on its own initiative.

The personalizing of the telephone has already begun with the rise of GSM. Mobile Internet will really bring personalized information and entertainment. The mobile terminal is a personal gadget. It gives the Internet user access to personal information wherever he or she might be. Much of the personal information will be configured via a fixed PC on the World Wide Web, but will be subsequently requested via a mobile terminal.

Financial transactions form a third aspect of mobile data communication that offers a substantial supplement to the current Web. The GSM terminal can offer important added value to electronic commerce, particularly concerning security aspects. The security mechanisms in the mobile terminal (for instance, SIM card) will play an important role in identifying the user. Data to be sent through the ether can be encoded. Payment authorization can take place via communication with a server. From now on transactions can take place anywhere: on the street next to the parking meter, by the soda dispenser, at the door when a parcel is delivered, and so on.

Due to these developments, Ericsson believes that by 2004 there will be at least 1 billion mobile network users. Of those, around 350 million will make use of mobile Internet services (see Figure 1.14). The market for mobile data services will therefore grow extremely fast in the coming years—even faster than the growth that we are now seeing in GSM and the Internet.

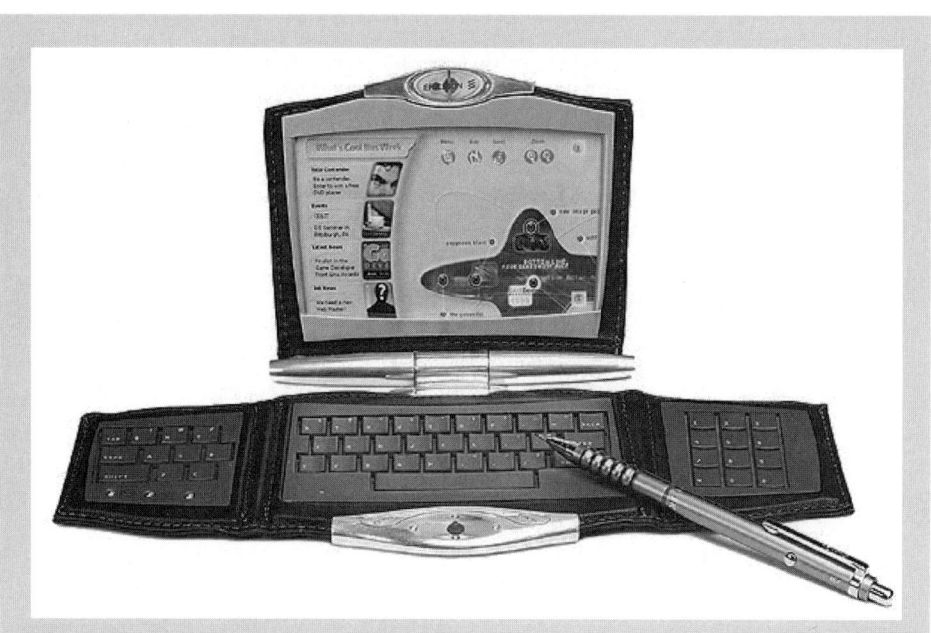

Figure 1.14
Prototype of UMTS terminal by Ericsson.

GPRS (General Packet Radio Services) is an exceedingly important technique in the evolution of mobile Internet services. GPRS offers the GSM network properties such as "always online," payment per kB, and—not to be forgotten—higher transmission speeds. The third-generation (3G) networks such as UMTS (Universal Mobile Telecommunications System) will bring still higher speeds of up to 2 Mbps. With these speeds, sending pictures is no problem and a multitude of new uses will be available.

But wireless communication will go further. Ericsson's vision is that eventually all communication cables will be redundant. The global solution for this is Bluetooth. It is a short-distance radio technology that enables all types of apparatus to communicate with each other. The short reach of the Bluetooth technology (up to 10 yards, and in a later phase to 100 yards) is specially intended for household applications. Think in terms of wireless communication between thermostat and server, between PC and ASDL modem, and between keyboard and PC. Bluetooth was introduced quite recently. In 2005, it is estimated that 900 million devices (telephones, PCs, household equipment) will be equipped with a Bluetooth chip.

> Mobile Internet is very much more than "cable-free" Internet. The freedom that people experience with GSM on the one hand and with Internet on the other will increase significantly after the two are merged. Applications that save the busy businessman 15 minutes every day are of immense value. And that is equally valid for applications that bring fun and entertainment to the masses. Ericsson's core activities are the development and building of networks and terminals. Ericsson has decided to move forward from its leading position in GSM to the 3G world and therefore is investing a large slice of its R&D budget on the development of WAP, GPRS, UMTS, and Bluetooth. But it will be the applications that make new networks and terminals necessary. Therefore, Ericsson is stimulating the development of applications through their own development and through diverse collaborations with content providers and software suppliers. The mobile terminal will be the access to the worldwide Internet, and that is the core of "the power of mobility."

NEW POSSIBILITIES

Positioning

Many services can deliver extra convenience and added value to the user if the location of the user is known. Think, for example, about maps and routes to the nearest gas station, bathroom, McDonald's, or ATM. In addition, it can be relevant for users to know where others are located. How often do you end up looking for someone in a public place like an airport, gas station, restaurant, or hall? In the United States, positioning has been made compulsory for mobile operators to support the alarm number 911.

GSM networks are not developed to give locations of mobile telephones for use in services. In a GSM network, it is possible to crudely determine where a mobile telephone is located. This method is called cell-ID. The location can be determined most accurately during a call. During a call the location can be established with an accuracy of between 200 yards and 3 miles. The accuracy depends on the cross-section of the GSM cell where the user is located. A GSM cell is the reach of a GSM mast. The cells are at their smallest in those areas where telephone traffic is most dense, for example, around traffic intersections and in densely populated cities. The reach of a GSM mast and thus the size of a cell vary due to atmospheric conditions. When the telephone is not in use but is switched on, the location can be determined at the location area level. In addition to cell-ID, a number of technologies for positioning are already available or in development: Global Positioning System (GPS), Time of Arrival (TOA), Time Difference of Arrival (TDOA), and Enhanced Observed Time Difference (E-OTD).

Global Positioning System

GPS is a position-determining system that was developed for military purposes by order of the U.S. Department of Defense (see Figure 1.15). GPS has an accurate version for authorized users (Precise Positioning System) and a version available for everyone (Standard Positioning System). SPS contains an intentionally added fault, making it less accurate. Differential GPS was developed to correct this defect. The U.S. Department of Defense can switch off GPS at any time. GPS is very accurate. The accuracy depends on a number of factors, such as the angle between the satellites, measured from the point from where the location must be determined. Roughly, for Normal GPS (SPS) in 98 percent of cases, accuracy is within 100 yards and in 75 percent within 50 yards. The accurate GPS version (PPS) is accurate to within 25 yards. Differential GPS is accurate to within 5 yards. The intentional fault was taken out on May 1, 2000, improving accuracy to within a few yards.

Figure 1.15 Combined GPS and GSM module designed for the Handspring Visor.

A disadvantage of GPS is that a clear path is needed to at least three satellites and this can be a problem in a city. Interference caused by PCs and TVs can also influence the accuracy. A last disadvantage is that a minimum of 10 seconds and up to half an hour is needed to fix the position, depending on when the device concerned last used GPS. GPS can be combined with a mobile telephone by building GPS into the device or by taking up D-GPS in the network. Due to the costs of the equipment and customer-specific requirements, the combination is not yet available for the mass market. GSM and GPS systems are often built into delivery trucks or automobiles to establish the vehicle's location in the case of theft or for navigation.

Triangulation

There are a number of techniques that make use of GSM for positioning. These are known as triangulation because the measurements of these signals usually take place from three angles. The best known are Time of Arrival (TOA), Time Difference of Arrival (TDOA), and Enhanced Observed Time Difference (E-OTD). TOA measures how long it takes a signal to travel from the base station to the mobile telephone or vice versa. Receivers must be placed on the base station for this. The system works on all mobile telephones and the location can be found very quickly. If more base stations measure the TOA of the same cellular telephone, the resulting time differences can be used to calculate differences in distance. TDOA is based on this. Synchronization of the base stations is essential for this to work. In the case of E-OTD, the mobile telephone measures signals coming from three different base stations. Also here, the base stations must be synchronized or extra equipment must be installed. Nokia has determined that synchronization is preferable. The location calculations are carried out by this equipment or by the cellular telephone. In the first case, an adaptation to the software in the mobile telephone is sufficient. In the second case, the mobile telephone's processor capability must be expanded. Advantages of this technology are the low load imposed on the GSM network and the fact that the location is also available when the device is just switched on. Disadvantages are the adjustments needed to both devices and network, meaning that existing devices cannot make use of this technique. The synchronization required slows down the positioning, leading to mistakes if the device is moving at a high speed through the network. All techniques mentioned are hindered by reflections against buildings. This causes the signal to remain longer in transit, leading to incorrect location determination. Specific algorithms can eliminate the effects of reflections. With these, an accuracy of 125 yards is possible in 67 percent of cases.

0804 Measurements Report

Cellpoint has developed an application for the mobile telephone that requires no alterations to the GSM network. The application makes use of an existing management report. This report is sent when the line is open and in other situations it is saved on the SIM card. These reports actually end up in the wrong place in the GSM network, causing the information to be not easily available to the service.

Each technology must be judged on the additional investment required in the network, alterations needed to the mobile telephone, and the required accuracy. Within the telecom world, both operators and suppliers are working on the standardization of positioning technology. The demand from the U.S. Federal Communications Commission (responsible for the distribution of licenses and the award of frequencies in the United States) to make positioning compulsory to support the 911 alarm number service is a key factor here. Also, the possibilities of offering commercial services will be taken into account.

Without yet being able to say which supplier will provide the dominant positioning technology, it is clear that positioning is a relevant development to support the mobility of the user, with services specifically developed for mobile use. This is not an option in the normal Internet and is unlikely to quickly become so. Quite apart from the technological developments, user acceptance is an important condition. Positioning can produce numerous new applications that can offer the user benefits and convenience. Still, users will probably be discerning in their choice of service supplier and go for the most trustworthy option. Also, legislation mandates conditions about the uses of positioning. Later in this book, privacy aspects are covered in detail.

Location-based services are seen as one of the important revenue generators in the 3G era. Building the right applications and finding the way to offer these services commercially in a successful way may take some time. Customer acceptance is another important factor. Therefore, most networks will implement location-based services before the introduction of 3G services. The combination of GPRS and location information can offer application developers enough opportunities to build and market location-based services. The different implementations might restrict the offer to the home network operators. Customers using other networks—while on vacation or on a business trip—might find it difficult to use these services, because their own operator might not be able to offer them outside the home network. These customers must find and use the services offered by local operators, sometimes in a different language. Location-roaming services will be the next challenge.

Synchronization

According to Durlacher Research, synchronization is the process of maintaining identical applications or data where the user wants it. A calendar application user will want to regularly synchronize the calendar on his or her Personal Digital Assistant (PDA) or cellular phone with the calendar on his or her company network, so that the information is always up to date. Synchronization is crucial for mobile Internet, because there will be a need for both Web-based and local applications on a PC or a mobile device. The limited penetration of PDAs in the business environment has given the lucky owner of a PDA synchronization problems up to now. As soon as increased productivity is a certainty, companies will equip many employees with PDAs. The PDAs will then have access to intranet, enterprise resource planning, customer relationship management applications, and the current Microsoft or Lotus office environment. Such applications require the presence of local data. The absence of mobile coverage will be an argument for synchronization in many countries. Within buildings and in isolated areas, synchronization will be important. Also, business travelers who work regularly during their flights will need to synchronize their calendar, email, and files after arrival.

Speech Technology

At first glance, speech technology appears to be a competitive development. This is because the user communicates by speech instead of via a screen. The reverse is really true. When it is possible to approach a central database in various different ways, the value of the whole will only increase. Manufacturers are developing software that will enable information from the same XML source file to be accessed via speech as well as via WAP. The World Wide Web Consortium has accepted VoiceXML 1.0 as a standard for speech-controlled Internet use. This will enable the application of interactive voice response systems on the Internet to rapidly gain momentum. Applications are being developed that will convert speech and pictures. These speech technologies will certainly be combined with WAP in the near future. Chapter 2 deals with this in more detail.

Bluetooth

Bluetooth is a technology for wireless communication. The Bluetooth consortium was set up with the aim of enabling many types of apparatus to communicate with each other via wireless means. Bluetooth is intended for a wide range of computing and telecommunications devices, such as PDAs, laptops, cellular telephones, and cameras, but also printers, faxes, and video recorders. Bluetooth promises connectivity without the need for cables or proprietary software interfaces. The devices must not be more than 10 yards apart. Using an amplifier, this reach can be increased to 100 yards. Bluetooth has been developed for speech and data. The speed will increase over the coming years, also making transport of video possible.

Ericsson invented Bluetooth technology. In 1998, Ericsson, together with IBM, Intel, Nokia, and Toshiba, set up the Bluetooth Special Interest Group to further develop Bluetooth. Version 1.0 of the standard was set down in 1999. Ericsson has made the technology available license-free to develop a wider market than just mobile telephony. In the meantime, more than 2,000 suppliers have said that they will make use of this technology.

Bluetooth will be a handy solution for numerous situations. For example, a cellular telephone equipped with Bluetooth will be able to call out at home via the normal telephone line (lower tariffs). En route it can be used via the mobile network and will be able to operate as a sort of walkie-talkie if it is within reach of another Bluetooth-equipped device. Teachers have a lot to look forward to in the future, as pupils will be able to whisper not only to those sitting next to them, but to everyone in the class.

Bluetooth is also a welcome addition for computer users. With a laptop you will be able to reach the Internet from anywhere, regardless of whether you use a mobile telephone, the LAN, or your PSTN/ISDN or ADSL connection. Users can go with what is easiest and cheapest at the time. During a meeting, participants with a laptop or PDA will be able to exchange documents and electronic business cards automatically. Also, the perpetual irritation of having to synchronize all types of electronic calendars will be history. The desktop computer, the mobile telephone, the PDA, and the laptop will synchronize automatically once they are within reach of each other. The combination of Bluetooth with WAP makes a cellular telephone a universal remote control. WAP applications for granting access and controlling all types of apparatus will be within reach.

In 2000, Ericsson introduced Bluetooth accessories because they did not expect that users would replace their GSM telephones or PDA right away. A big advantage of Bluetooth compared to infrared is that an unbroken line of sight is no longer needed. The organizer in the hand can communicate with the mobile telephone in the briefcase. Much is expected of the wireless headset (see Figure 1.16). This can be connected to the cellular telephone in your back pocket or purse or to a normal telephone. Thus, you will be able to keep your hands free for driving, writing, or other activities. Also, payment in shops could be easier when your cellular phone can act as a digital billfold. Ericsson is also expecting much from their Cordless Screenphone (HS210), a writing pad with a touchscreen.

Figure 1.16
Ericsson Bluetooth headset and phone.

Bluetooth faces competition from the following technologies:

- Microsoft's Universal Plug and Play (UPnP) supported by 3Com, HP, GE, and IBM is a competitive standard that is also aimed at PCs, PC peripherals, PDAs, and cellular telephones.
- Jini from Sun, Sony, Cisco, Motorola, and Oracle. Jini is based on Java.
- HAVi (Home Audio/Video interoperability), supported by Sony, Philips, Sun, Pioneer, and Sharp. The focus of these technologies is interoperability between digital audio and video apparatus such as cable modems, set-top speakers, and Internet televisions.

The division between these groups and the contribution that many of the supporters of competing technologies give to Bluetooth suggests that, although the battle is not won, Bluetooth has a good chance of evolving into the standard.

The expectation was that the first devices with Bluetooth would appear in the market in the spring of 2000, as extensions to existing mobile telephones, notebooks, or digital cameras. Intel was to deliver the first Bluetooth chips at the end of 1999, but delays in delivery forced suppliers to come onto the market later. The price of the Bluetooth chip was still very high at the end of 1999 ($35) and $25 near the end of 2000. The chip must eventually cost around $5 to enable production of affordable Bluetooth devices. The risk is, of course, that a chicken-and-egg situation will arise, because the price will only go lower when a specific production quantity has been achieved and this will not be reached because of the high price. Another old standard for wireless communication is currently implemented by some notebook producers: 802.11B. Apple has implemented this standard in most of its products because it is cheaper, more reliable, and can be used over longer distances.

In spite of this, research bureau IDC expects that in 2005 some 2.8 billion devices worldwide will be equipped with Bluetooth, split evenly between mobile devices and "fixed" apparatus such as computers, faxes, videos, and printers. Ericsson estimates that in 2002, 100 million mobile telephones will be equipped with Bluetooth.

ALWAYS AND EVERYWHERE

At the beginning of this chapter, the question was posed whether, now that the first WAP services have been introduced, is this hype or a revolutionary new technique. When and how you will have access to information over the whole world meets a fundamental demand. Still, WAP is not the first facility that enables access to information at all times and locations. Laptops with a GSM connection or a special communications device have offered this for a while. Why will WAP, in combination with GPRS, be a success, whereas the earlier-named access techniques were not?

Standards

The main advantage of WAP is that it is a standard that is supported by all parties in the chain, from mobile customer to information, and it is wholly developed for mobile communication. Mobile equipment suppliers such as Nokia, Ericsson, Motorola, Samsung, Alcatel, and many others offer devices with a WAP browser. All mobile operators have launched a WAP service or are working hard to introduce one. In addition, there are also other parties, independent of the mobile operators, such as banks and portals, who offer services via their WAP gateways. It is relatively simple for providers of information and services to make services using WML. It is also important for these parties that their information or services are available to a wide group of users. With the wide support for the standard and the larger potential target group (not only users with an expensive laptop or communicator), it seems that this will be possible with WAP. The initial scarcity of WAP devices is also over, enabling the target group to get hold of the necessary devices. We expect that with the forecast growth in bandwidth, a need for more functionality within WAP will arise, particularly in the graphics area, for example, color and moving pictures. It is expected that the WAP standard will be extended to accommodate this in the future. The industry might also decide to move to another markup language, such as cHTML. As long as the standard is continuously supported by the whole industry, an important barrier to the growth of mobile Internet will be removed. After all, uncertainty about standards leads to much wasted energy by manufacturers and hesitance among consumers.

Sharpened Service Offerings

In contrast to other means of mobile access to the Internet, WAP services are focused on the mobile situation. Service suppliers make their products especially for the mobile situation, thus taking into account the limited bandwidth and the limitations of the cellular telephone. The user receives faster access to the information that is relevant for him or her.

Complementary Technologies and Possibilities

Because it is an open standard, WAP will benefit from the development of complementary technologies and facilities. Compared to WAP, SMS is an inferior user interface. As a push medium, SMS can give users the option to have their attention drawn to files, new email, appointments, stock prices, and sports results. Technically even offers from a supermarket that the user is driving past are possible via SMS. The question is, does everyone want that? In this respect, SMS is supplementary to WAP. The SIM toolkit is complementary in that it can provide end-to-end security, enabling the user to make simpler and safer use of M-commerce. The combination of WAP with faster carrier services, such as GPRS and later UMTS, will increase user friendliness and the speed of information transfer. The integration of positioning and alliances like Symbian and Bluetooth will cause an enormous increase in the number of available applications and ease of use. Positioning will stimulate the personalizing of services. Symbian will speed the development of mobile Internet handsets. Bluetooth will make interaction with nearby equipment and use of a mobile telephone as a universal remote control possible. The more applications that become available, the faster that both the use and the number of users will grow.

2 Similarities and Differences between WAP and the Internet

In this chapter...

- WAP versus the Internet for the User   46
- WAP versus the Internet for the Service Provider   61
- WAP versus the Internet for the Service Developer   70
- WAP and Corporate Networks   77

In the media, WAP is presented as the mobile Internet. This explanation has different interpretations. Can the World Wide Web, with all its pictures, sounds, and animation, be accessed from a mobile phone? Or is it like sending email from your laptop? Can you download music and listen to it via your mobile phone? The term "mobile Internet" brings expectations that cannot be fulfilled with the first versions of WAP. This leads to unnecessary disappointment, as access to the worldwide information transport network can offer much more. As mentioned in Chapter 1, there are significant functional differences between a mobile phone and a PC. A person will use a mobile phone in different situations, for different reasons, in different places, and at different times than the Internet.

The differences between WAP and the Internet via the PC have substantial consequences for the suppliers of services via WAP. It has often been said: "I already have a beautiful Web site. If that Web site is simplified and made accessible for WAP, I am ready." Copying existing World Wide Web services to WAP is relatively simple in many cases. Web site programmers can program a WML site without too much trouble. It is unlikely that the new possibilities WAP offers can be exploited to the fullest and lead to satisfaction for the user. When the Internet gained popularity, any self-respecting company wanted to be present on the Net, often without having an objective in mind. Most of the sites were nothing more than outdated electronic leaflets without any added value. The comparison with the introduction of WAP is striking. The big difference is that entrepreneurs, who have completed the learning curve for the Internet, can now complete the WAP learning curve faster. In this chapter, we elaborate on the similarities and differences between WAP and Internet via the PC, from the perspectives of the user, provider, and developer.

Some readers might wonder, "Why pay so much attention to the similarities and differences? Do both sides not move toward each other and will we not have devices that have all features: telephony, Internet access, and television?" We do not believe in the theory that people will only need one device in the future. We prefer to believe that people will have more devices, for specific moments and specific uses. As a result, service developers will have to keep in mind the fundamental differences between services aimed at mobile users and those aimed at PC users. To clarify the differences as much as possible, we have chosen to discuss WAP and the Internet via the PC in greater detail. In the following chapters, we will use the term "mobile Internet" more when we discuss the market and trends in that market.

WAP VERSUS THE INTERNET FOR THE USER..........

Lately, Internet users have gotten used to the never-ending variety of Internet services. Uniform use and worldwide access to a wide array of services have made for a rapid increase in Internet use. As explained in Chapter 1, WAP potentially offers the same success factors for a mobile Internet. The opportunity is there for an unlimited amount of services to become available via WAP. These will most likely not be the same services as those available for the Internet today. There are plenty of differences between WAP and the Internet, requiring service adaptation.

Before we cover these differences, we would like to emphasize the similarities between the Internet via the PC and WAP. Users familiar with the Internet via the PC will soon realize that WAP also includes scrolling and clicking on hyperlinks. The previously educated user does not have to learn how to use a site with every new provider. Users surfing their way to a site will find their way around intuitively. Another similarity is the online connection to a worldwide network. WAP sites can be accessed in a uniform manner using a URL. These URLs can be imported and stored onto mobile devices, similar to the bookmark function on a PC Web browser. The importance of search engines is identical to that of the Internet. There has also been tremendous growth in the amount of WAP sites, both good and bad. Just as with the Internet, it is often difficult for the user to obtain good information and for the supplier to reach the end user. The limited user interface of the mobile phone makes it more difficult to find sites, so portals and search engines are therefore of great importance for mobile Internet users in finding a site.

Differences between Mobile Phones and the PC

The differences between the mobile phone and the PC are obvious. Ease of use and ergonomics of mobile phones will have to improve greatly to close the gap between the PC and the mobile phone. Linked technology developments such as speech recognition and hybrid devices are relevant in making the use of information services via mobile devices easier.

Ease of Use

Have you ever written an SMS message on your mobile phone? Have you ever had a fax or email sent to your mobile phone? Have you ever tried to find a specific function in the menu of your mobile phone? If you have experienced one of these activities, you know enough. It is not easy to enter text with 12 keys and a few soft keys. Reading substantial amounts of text on a 2-inch screen is not easy to do (see Figure 2.1). Functions can often not be found without consulting the manual. On top of this, the screen is often black and white and the information is slow to appear. The sluggish speed is a result of the limited memory capacity of the telephone and the limited speed of the connection. Keeping in mind that the battery of a mobile phone will allow 30 to 120 minutes of WAP time without recharging, we have our answer: Long live the Internet via the PC, and let's use the mobile phone to make and receive calls. End of story?

Figure 2.1
Small screen.

No! Mobile phone manufacturers are working on enhancing the user friendliness of the mobile phone, and the PC also has a number of drawbacks.

What are the most eye-catching differences?

- *Size and weight.* The mobile phone is much smaller than the PC and weighs much less. Even the smallest laptop still weighs about 1 kilo, and measures $8 \times 6 \times 0.81$ inches. The small WAP device from Sony, the CDM-Z5 (Figure 2.2), measures $3.5 \times 1.9 \times 0.85$ inches and weighs only 0.18 pounds.

- *Size of screen, resolution, and color.* A PC often has a 15-, 17-, or even 19-inch screen and a resolution of at least 1024×768 pixels. Most monitors also have about 64,000 to 16 million different shades of color. Most mobile devices have a screen measuring 1×1.4 and only black and white for color. The Nokia 7110 has a resolution of 90×65 pixels.

- *Processor capacity and memory.* Any PC that was bought recently has at least 64 MB of RAM, a processor speed of 600 MHz, and a 20 GB hard disk drive. The memory of most mobile phones stores 150 to 1,000 telephone numbers, and an additional 250 numbers can be stored on the 32k SIM card. Most mobiles use a 16/32-bit processor with a speed of about 13 to 20 MHz.

- *Keyboard and mouse.* A PC has a mouse and a QWERTY keyboard, often supplemented with other function keys and a numerical keyboard. This makes using a PC easier. Most mobile phones have 12 keys and two function keys. These keys are used for navigation and typing, usually by press-

Figure 2.2
The lightweight Sony WAP phone (CDM-Z5).

ing keys multiple times. Entering text on a mobile phone is therefore more complicated and time consuming than on a PC.

In short, the PC scores much better on most issues, except for size and weight. In this respect, the mobile phone is ahead of the PC. This mobility makes the addition of information services to the mobile phone very interesting. The mobile phone also has a SIM card, which can be used for transactions and personalization. We discuss these functions later. Finally, the price of a mobile phone is significantly lower than that of a PC.

What do mobile phone manufacturers do to simplify surfing on a mobile phone?

Ergonomics

Even though WAP has been specially developed for small screens and a limited number of keys, the device itself can be modified to enhance user friendliness. To make the screens easier to read, they now make up a larger part of the device than before and the resolution has been increased. Lately, devices with color screens have also become available. There are phones available on which the screen covers half the phone. Other devices have keys that appear on the screen that disappear when they are not used. Touching the screen controls these devices. The screen space can then be used for displaying other information. The increased resolution of the new generation of screens allows for more lines of text on a small screen without decreasing readability. The positive effect of the arrow keys and navi-rollers should not be underestimated. Using these aids allows for scrolling activity similar to that of a computer mouse through text on a monitor.

The limitations of the keyboard and menu navigation are the subject of much research. Ericsson has launched a mini-QWERTY keyboard measuring $4 \times 2 \times 0.5$ inches (Figure 2.3). This keyboard, called a chatboard, can be attached to the mobile phone and simplifies typing messages. For $50, users will not only get the keyboard, but also a mail address and a home page. A number of manufacturers develop soft key applications. The soft keys are those keys on top of the keyboard, serving different functions relevant for that moment. The function chosen will be displayed at the bottom of the screen. Soft keys often give access to pop-up menus with multiple options. Another improvement relating to software is the development of T9 by Tegic. T9 enables the user to enter text using a list of words. A number key only has to be hit once for a letter to appear. After pressing a number key for every letter, a word appears that is the most likely combination of letters pertaining to the number keys hit. If this is the correct word, the next word can be extracted from the list of words. The number of times a key needs to be hit to type a word is reduced considerably, making entering text easier and faster. T9 has been licensed to 90 percent of the mobile handset market.

Figure 2.3
The chatboard developed by Ericsson.

Japan has taken a jump start in the development of miniature devices for specific target markets. Even though Japanese operators use different network standards, the innovations with respect to device ergonomics will also be implemented in U.S. and European markets. Devices have been developed with Sky Melody. With this technology, the user can program a popular melody with MIDI quality via the Internet to substitute for the ringing tone. Telephones have been developed for children with mobile cartoon characters such as Hello Kitty and Tare Panda (see Figure 2.4). These characters jump across the screen and entertain children with questions and jokes. Developments in the Japanese telecommunications industry will be covered in greater detail in Chapter 5.

Figure 2.4
Examples of Japanese mobile phones.

Hybrid Devices

The Nokia 9000 Communicator was the first mobile phone with a QWERTY keyboard, a larger screen, and built-in data communication options such as SMS, email, and the Internet. Devices like this are called hybrid devices and are offered by all major device manufacturers. Even though the latest version of the Nokia Communicator and all its competitors are slimmer and lighter (see Figure 2.5), many users still find the size and weight to be a problem. They prefer a lightweight device with a longer standby time, which gives them access to a large number of services. Palm Pilots and

Figure 2.5
The Nokia 9210 communicator.

laptops make it easier to read larger documents. With these devices it is no problem to access the Internet with a regular HTML browser, certainly not after the introduction of the GPRS. Many people already own a laptop as well as a mobile phone. Even if laptops allow for faster data communication, mobile devices will remain on the scene. The user does not always want to carry a laptop around. The laptop or Palm Pilot will also be used to access Internet or intranet applications.

The hybrid mixes of mobile telephones and laptops will not replace the telephone in the near future. They will be purchased for specific tasks and in a specific situation at the same time as mobile phones. For example, it is handy to use a laptop as a notepad in a meeting but not to carry it around at all times. Hybrid solutions that make other devices unnecessary may be the exception. An obvious combination is that of electronic organizers with telephone functions. The electronic organizer replaces the calendar and address book and is therefore no ballast to carry around. A compact hybrid organizer with telephone functions will be well received by people already carrying around a day planner. Another example of a hybrid device for the younger target market is the combination of mobile phones with a Walkman. Besides listening to the radio, tapes or discs, music can be downloaded from the Internet. Instead of a Walkman, a combination with a car radio is also a logical development. Blaupunkt integrated a GSM car radio more than two years ago (see Figure 2.6). The installation of a GSM connection in the car radio makes downloading music possible and, in combination with a car kit, will even contribute to safety on the road. Car radios in combination with WAP offer new possibilities, like the use of destination locators via mobile networks. Sony Playstation 2 has already been outfitted with access to the Internet and even the simplest of phones have games preinstalled. What could be cooler than playing a game with friends even when you are not in the same room? It should not be surprising that there is great interest in WAP within the entertainment industry.

The hybrid mixes described are only examples of the assortment of mobile devices that will be developed in the next few years. It is clear that many people will have different devices for different situations. Palm, Microsoft, Ericsson, Siemens, and others supply PDAs requiring a mobile phone for an Internet connection. PDAs

Figure 2.6
Blaupunkt Monte Carlo, a combination car radio and GSM phone.

are discussed in greater detail in Chapter 5. In the future, the majority of mobile phone users still will have at least a lightweight mobile phone in their possession at all times. Larger mobile devices or devices with a connection to a mobile device will be owned by a large part of this population, too.

Voice Recognition

In the longer term, the integration of voice recognition will simplify the import of data to devices or services. Mobile operators have been offering network-based voice recognition services to access a personal address book since 1997. Almost every device manufacturer offers phones with the option to voice dial from a limited list of names. Voice recognition has had limited success in a mobile network environment, simply because the end user has to try too hard to teach the names to the system, as the technology is sensitive to background noises. The development of speech recognition has, however, caught the eye of technology giants such as Lucent, IBM, Philips, and Microsoft. These companies have invested heavily in research and they have acquired promising voice technology companies. The Belgian company Lernhaut & Hauspie, partly owned by Microsoft, is the market leader in the area of speech technology. Lernhaut & Hauspie's Voice Xpress version 4.0 can recognize every common language and supports Microsoft Office applications. In the United States, voice portals exist for the Internet that translate the sites into spoken language. A large voice portal, Tell.Me, publishes information on entertainment. The user calls a toll-free number and navigates via spoken commands. In combination with regular or mobile Internet, more options are possible: Instead of navigating through a voice-response menu, choices are displayed on a screen and the user can call out the desired option. Why not place an order verbally and confirm and pay via the screen of your mobile phone? Speech can be activated in two ways: questions and answers. In the case of questions, the words used have to be translated into commands by using voice recognition technology. This is not a simple process within the mobile environment, in which background noises are the cause of many mistakes. To offer a combination of text on the screen and voice for a wide array of services, voice recognition will have to be improved, especially in the area of larger applications for mobile devices. The transition from text to speech is easier to apply in a mobile environment. The problems with background noise are not relevant in this case. Besides the translation of Web sites to speech, reading emails has also become popular. The Van Dale (the Dutch equivalent of Webster's Dictionary) is developing an application for reading email in Dutch, and a separate module can be activated for the English phrases. The market for voice recognition applications currently amounts to $2 billion worldwide. Analysts expect an annual growth of 20 percent for the next few years, mainly in the area of services developed for mobile phones.

Telephony Services

The example just mentioned, touches on another difference between WAP and the Internet, namely, the relationship with telephony. Internet and telephony have hardly been integrated. From a user's perspective, a PC has not yet become a telephone. Sites or applications combining the Internet and telephony are rare. Privacom, with Call me Now! (the 1-800 number for the Internet), is one of the few suppliers of telephone services. The opposite is true for mobile telephony: A mobile phone is primarily bought to make calls. Calling from a WAP site will not only be easier for the user, it will be done without giving it a second thought. Scoot, the telephone directory, and the Dutch Yellow Pages have realized this and are offering WAP access to their services. The user searches for a number and uses the mobile phone menu to call the number. The second version of the WAP standard uses the Wireless Telephony Application to standardize the integration of telephone services, making its use easier.

Availability

The mobile phone offers something that a PC does not: the possibility to call from virtually anywhere as long as there is sufficient network coverage. Coverage has improved enormously in virtually every country, resulting in accessibility from almost any area. Most networks offer good indoor coverage, especially in urban areas. This means that a mobile phone user can access a network from almost every indoor and outdoor location. For WAP applications, this means that people can inquire after information via their mobile phones. The mobile phone is not solely for use on the road. Many people do not have an Internet connection at their disposal, but do have a mobile phone. A television watcher responding to a TV show or participating in a game show will use the mobile phone. The PC is in the study and the waiting line for the toll-free number is often too long.

Threshold of Purchase

Employers will not always provide Internet access at work. Most employers cannot and will not facilitate Internet access for all employees. The threshold for availability of Internet access is high. This is due to the high fixed cost of PCs, Internet subscriptions, dial-up costs, and similar costs for space, furniture, and cables. For many European companies, the Internet is still in the experimental phase, in which personal use of the Internet during work hours plays an important part. The purchase of a PC for home use is a substantial investment. Many people underestimate the cost of purchase and use of a PC. They do not realize that a PC generally needs to be replaced with a newer model after two years. Finally, there is still a lack of knowledge and insight on how to operate a computer, let alone install new programs.

The purchase threshold for a mobile phone on the other hand is low. A telephone is not expensive and is relatively easy to use. WAP will become a standard part of every mobile phone. Because almost every employee has a mobile phone at his or her disposal, paid for by the employer or not, they can be reached. As with the Internet, personal use of the mobile phone is a problem, but to a lesser extent. This is because decision makers often understand more about mobile telephony than about the Internet. "The Boss" has had a mobile phone for years and sees the efficiency advantages a mobile phone brings. Moreover, employees consider a company mobile phone to be a fringe benefit. Companies also have the ability to control the use of telephones by blocking numbers and evaluating management reports. Larger companies will probably block WAP access numbers to limit WAP access of their employees to those deemed relevant.

User Threshold

Another advantage WAP has over the Internet is the lower user threshold. Many questions, ideas, and activities can appear out of the blue and at any time, any place. A great number of people sleep with a notepad on the nightstand in case they need to remind themselves of something. It used to be that people would tie a knot in their handkerchief as a reminder. Mobile phones offer prompt solutions with regard to calling people. Why not order or inquire after something when the thought strikes you? Besides the previously mentioned location-independent character of the mobile phone, the mobile phone is easier to access than a PC. The mobile phone is not only always within reach, it is also always switched on. Dialing into WAP takes about 15 seconds with a circuit-switched data connection. Using packet-switched networks like GPRS reduces dial-up time to near zero. In the case of a PC, the Internet can be accessed with ISDN, ADSL, or cable. Starting up a PC takes about two minutes, unless it is already switched on. As WAP is available instantly, a user will utilize the WAP services at moments he or she is waiting or is bored and in need of distraction. Searching for a joke while standing in line at the grocery checkout, checking email while attending a conference, and evaluating stocks just before a meeting are situations we find ourselves in several times a day.

Way of Use

The situation for mobile phone users also differs from that of PC Internet users. In using the Internet, a PC user will sit behind and look at a screen. All attention is aimed at the screen. WAP use often takes place in a dynamic environment in which a user is multitasking. Therefore, not that much attention is paid to the screen.

WAP is used for different activities. For extensive searching for product information, vacation destinations, or encyclopedias, there are better tools then a mobile phone. WAP services, which offer large pieces of text with limited navigation options, will receive no support from users. Detailed, up-to-date information such as news headlines, Dow Jones closing rates, and email is relevant to many people. As a result, the demand for services focused on mobile phone usage will increase. Obvious application choices that are already in existence are ATMs, restaurant listings, or hotel locators. Where the Internet is suitable for in-depth searches for information, detailed graphical pictures, and detailed drawings, WAP is all about condensed messages, direct answers to questions, and urgent and up-to-date information. Even if there will be more technical capabilities for WAP, the information a user looks for via a mobile phone will be different from the information that same user will search for on a PC. Mobile use of services is mainly ad hoc timely use, multiple times a day.

Number of Users per Connection

Another difference between WAP and the Internet is the number of users per connection. A mobile telephone connection almost always has one user. A number of years ago mobile phones were often shared both in a business and personal setting. The decrease in cost of mobile telephony has ended this. Many users often share an Internet connection. This occurs both in a business and personal setting. A single-user connection in the workplace has both positive and negative aspects for users. Users can be held directly responsible for their behavior by their employer. This way the cost aspect and the type of sites visited can be controlled better. Similar to mobile phones, an employer can expect an employee using WAP to be available at all times. An example of this is traffic information. As employees now have access to traffic information, they can be expected to circumvent congested areas whenever possible. The advantage of personal access to the Internet via WAP is that the user or his or her employer can indicate what information is relevant. This makes personalization possible, simplifying and streamlining the searching process. A user will be offered a personalized menu.

Perception of Security

Personalized access makes payment via WAP much simpler. The owner of the SIM card in the WAP telephone basically has an ATM at his or her disposal. The perception of security of mobile telephony and the Internet differs. Even though there are various methods of payment for Internet purchases, the perception of security is still an issue. The mobile phone is generally regarded as secure. The expectation is that users will use their mobile phones en masse for purchasing products and services. It is not unlikely that the mobile phone will be used to pay in stores or to pay for parking. Various mobile operators experimented with vending machines, allowing the user to order and pay for a product with his or her mobile phone. The payment for low-value products can occur via the mobile phone bill or the user's prepaid account.

Methods of Payment

Payments via the mobile phone will not only increase in number as a result of the perception of security, but also because of the available methods of payment.

Even though access via the PC is always paid for, regardless if this is per month or per minute, payment for items purchased on the Internet through a bank account has not yet been developed. This limits the possibilities of small payments. With mobile telephony, this is different. From day one there is a relationship between the WAP user and the network operator. This relationship has been used at @info to bill paid content for information suppliers via the mobile invoice. A telecommunications operator has the choice of becoming a bank. Zed, part of the Finnish Telecom company Sonera, has expressed this ambition. It must be mentioned that the concept of a mobile invoice (or the mobile prepaid amount) as payment for mobile Internet has a number of disadvantages. The average user is not used to a telephone bill amounting to hundreds or even thousands of dollars. As a result, most mobile operators limit the amount per transaction at the moment. Employers also have second thoughts about mobile phones being used as company credit cards. Some telecommunication operators have therefore started alliances with banks or are acquiring banks. Making payments via mobile phones against regular bank accounts or credit cards is an attractive alternative. Again, the relationship between user and mobile operator contributes to simpler and more secure methods of payment.

Price Sensitivity

Price sensitivity between WAP and the Internet differs too. The Internet is often associated with "free." Service providers try to set a price for their services. Access to restricted sites costs money in the form of monthly payments or a membership fee. Think about the restricted parts of bank sites, background information of news providers like the *Wall Street Journal*, and the files containing research for consumer affairs. Some other services that were free in the past will no longer be free as soon as it is technically possible to charge visitors. Mobile telephony was regarded as expensive a few years ago. The entry of new suppliers to the market has not only lowered prices, but the perception of the level of the prices as well. Calling from a mobile phone is no longer expensive, but it is not free either. Mobile Internet via WAP can therefore be charged. The rates will have to be determined based on the added value. The added value for the user depends on the situation that person is in at that point in time, and of course on the quality of service.

Integration of WAP and the Internet

In our opinion, the keyword for the success of WAP services will be integration. Integration of telephone services and WAP services, integration of different functions in one device, and integration of other technologies such as T9 and voice recognition will contribute to improved services. The integration we are concerned with here is that of WAP services and services that can be offered via the Internet. Service providers can bundle the strength of different media, because the information can come from the same source. The information that can be found on the *De Telegraaf* (the biggest Dutch newspaper) Web site, is basically the same as the information on the WAP site for *De Telegraaf*. Customers can find a service provider with trusted services, regardless of the medium that a customer has at his or her disposal at that time. The services provided will be better when a Web site and a WAP site are combined with SMS. A financial supplier of services warns the customer via email and SMS if something unexpected happens that is relevant to the customer. The customer is on the road, not located behind the PC, and will look for more details via his or her mobile phone. Next, the customer will react, telling the service provider via the Web site, WAP, or a phone call, what the course of action will be.

Unified Messaging

Suppliers of services with added value have offered solutions to receive, store, and retrieve messages from different media sources for a number of years now. Picture, for example, voice mail messages as WAV files or receiving faxes in your email inbox, email alert via SMS, reading email via SMS, or reading email messages and faxes. In technical terms, this is referred to as unified messaging. Telecommunication operators have mainly offered voice mail services. Lately, the demand for an email alert function on mobile phones has increased.

The most important reason for the limited supply of unified messaging services is the complexity of the user interface. Voice response was the only media widely available for telephone users. The fact that unified messaging was limited to a select group of professionals until recently is also a relevant factor in its slow growth. Many busy professionals have secretaries or have no time, let alone the urge, to figure it all out. The fast penetration of the market of mobile telephony and Internet has resulted in a steady growth in a larger market. These days everybody has voice mail on a mobile phone, voice mail or an answering machine at home, and at least one email account. The Internet has taken away the complexity of the user interface to an extent. Xoip is a Dutch startup company offering unified messaging services via the Web, like receiving voice mail and faxes as email files. The number of services offered by Xoip is growing at a steady pace. This is the correct strategy, as simultaneous offering of a full range of services increases complexity and scares away potential users. Message4u, among others, offers services in which email can be read to the user over the phone. The Radati Group expects global market growth for unified messaging from $110 million in 1999 to $1.2 billion in 2003. This growth is closely related to the growth of mobile telephony and will have to result in a considerably more efficient use of time in the business world. The question remains whether the user will acknowledge these advantages and want to go through the trouble of incorporating the services into his or her daily routine.

Telecommunication operators will have to develop communication portals that offer access to telephony-related services from mobile phones, but also from the Internet (see Table 2.1). The launch of WAP will expedite the launch of unified messaging services for mobile operators. WAP solves the complexity of the user interface for mobile use to a large extent. Comverse, a large U.S. supplier of unified messaging services, has offered WAP voice mail access for two years now. The addition of WAP to voice mail is almost the same as the transition from vinyl albums to CDs in the eyes of the user. Whereas in the past the user had to listen to all the voice mail messages in the order in which they were received, this can now be different. Via WAP, the user can look at a list of voice mail messages. Using the telephone number or the name of the sender, the user can determine the priority of the messages and the order in which they will be retrieved.

Table 2.1 Overview

	Mobile Internet	**Internet**
Ease of use and ergonomics of the mobile phone and the PC	A mobile device is smaller and lighter than a laptop or PC. A mobile device has a smaller screen with lower resolution and a smaller limited keyboard. Transmission speeds are lower and network connection is more error-sensitive than with a PC with a regular Internet connection. There are high expectations in the area of miniaturization, device ergonomics, speech recognition, and the development of hybrid devices.	The PC with Internet access is superior to the mobile phone on a functional level. A PC has options like stereo sound, color, graphics, higher resolution and larger screen size, larger QWERTY keyboard, and faster, more reliable network connections. PCs and laptops are less manageable due to their weight and size than a mobile telephone.
Telephone services	Integration with telephone services is logical.	Integration with telephone services is in an early stage.

Table 2.1 Overview (Continued)

	Mobile Internet	Internet
Availability	Always and everywhere, even when you are on the road.	You can only work in a limited number of places in the house or at the office.
Barrier to purchasing	Low. Cheap and easy to install and to use. The fear of abuse can be treated by control mechanisms set up by the employer.	Higher. More expensive than mobile telephone, installation is more complicated, fear of abuse lies with the employer.
Barrier to use	Low. Always switched on, dialing in with two clicks.	Higher. PC has to be in the area, switched on, and available.
Way of use	Dynamic surroundings result in decreased attention of the user. The user will perform a targeted search.	Focus is on the screen as the user is seated behind the PC and has time to perform extensive searches via the Web.
Number of users per connection	Personal.	Often shared with family members or colleagues.
Perception of security	Perception of security results in payments via the mobile phone.	Is viewed as insecure, resulting in apprehension toward transaction.
Possibilities for payment	Various possibilities trusted by the user including micropayments.	Possibilities increase due to credit cards.
Price awareness	Mobile was expensive to use but will become cheaper. Relevant content can be charged.	Internet is often regarded as free, even if this is not really the case.
Integration possibilities	Specific services for mobile use, set up via Internet. Unified messaging services.	Specific services for use via the Internet, warning messages via the mobile phone.

WAP VERSUS THE INTERNET FOR THE SERVICE PROVIDER ....

The differences between WAP and the Internet do not only have consequences for the user, but also for the supplier of the services. It is obvious that carbon-copying Web sites to WAP does not do justice to the unique features of WAP, concerning the limitations, the size of the screen, the speed of data transfer, and the new possibilities for mobility WAP has to offer. In this section, we discuss the differences between WAP and the Internet from the perspective of the user, and translate this to the consequences these differences will have for the service provider.

Characteristics of a Mobile Phone

The visual differences between mobile telephones and PCs are clear. Whether it is screen size, keyboard, sound, graphics, or color, the PC offers more possibilities on all aspects except for mobility. The capabilities of PCs are greater due to a faster processor speed, longer battery time, and larger memory capacity. This means that the supplier of a WAP service has to keep these limitations in mind while determining its objectives for mobile Internet. Presenting large amounts of information or detailed blueprints is not desirable due to the limited screen size, lower resolution, and lower network speed. Transferring substantial computer programs, used to calculate invoices, for example, from laptops to mobile phones is possible as long as the number of input parameters is limited. The calculation takes place on a central computer and only the import parameters and the results are sent via the network. Offering services for which much data has to be entered via the mobile phone is not advisable due to the limitations of the keyboard.

The large variety among the different types of mobile phones with regard to size of the screens, microbrowsers, and keyboard functionality presents the service provider with an important choice: How far do you want to go in supporting all possible devices to reach the potential market? When we introduced @info, only one WAP phone model was available, the Nokia 7110. From the beginning the expansion of the product line was discussed with information suppliers. KPN Mobile intends to support, like every operator, all common models of the large suppliers.

Interoperability of the various browsers is improved in WAP 2.0. The device suppliers have no interest in the production of a device that receives limited support from certain gateways and is not suitable for visiting a large number of sites. Apart from all these problems, the expectation remains that the services can be accessed via different types of devices, first because the screen size will never be standardized, and second because there will always be differences between browsers of different brands and different versions. This is no different than the Internet, where there are many versions of Netscape Navigator and Microsoft Explorer and there are many plug-ins such as Flash, Acrobat, and RealAudio. Finally, because ease of use will be one of the most important differentiating factors of a WAP service, many different user interfaces will be developed for different terminals. For applications developed for a specific target market, the impact can be limited. Companies, for example, might supply their employees with only one type of mobile device. The intracompany applications can then be optimized for this mobile device.

For the development of a WAP service, the hardware a user has at his or her disposal needs to be taken into account. For very simple services, for example, three news messages that can be read, we will not have to develop a different user interface for every terminal. The service only uses the links and text and this is part of the minimum set that every WAP browser and WAP gateway have to support. As soon as the services use graphics, the size of the screen becomes important. The choice will then have to be made if the device with the smallest screen will be used or if the pictures will be displayed independent of the browser type. In the header of every request for information, the device type is indicated so that the correct setup can be forwarded. This could also be a menu option in the service, which the user can select if he or she wants to see a small or a large map.

Speed

Speed is important because the way mobile Internet is used requires speed. Speed is determined by the speed of the network, the peripheral equipment, and the services. The first two are determined by the infrastructure suppliers and the operators. Speed also has to be an objective in the design of the service. This brings about the role of the service provider. To use a service quickly, the visitor to a site must have a clear objective. From this we can determine which topics are searched most. Next, the service developer has to make those frequently visited topics quickly accessible.

Up-to-Date Information

Suppliers of news are forced to keep their site up-to-date at all times, as users do not usually visit twice a week, but twice a day. Users expect new information every time. If no new information is found, the frequency of use will decrease. The "refresh frequency" will determine the frequency of use for many information providers. Providing up-to-date information is a must for certain services. As a user of a sports information service, you expect to find scores during halftime, including the names of the players that scored. If these expectations are not met, it is unlikely you will visit this site again. The demand for current information is due to the nature of mobile telephony. Users expect current information, as that is the advantage of mobile telephony. Otherwise, they might as well just read the newspaper or wait until they get home so that they can watch the news on television. Whether you provide up-to-date information will determine the price the user is willing to pay. In the financial industry, it is common that the real-time stock exchange information is much more expensive than the same information on a 15-minute delay. The different rates the financial news suppliers charge for WAP information also indicates that users expect updated information.

Integration with Telephony Services

Users with a WAP telephone, a small Internet terminal, and a mobile telephone at their disposal create possibilities for a combination of WAP with telephone services. Standardized interaction between WAP and telephone services is possible with versions 1.2 and up. Many current devices have the ability to interpret a range of numbers as a telephone number and dial that number by pressing a few buttons. Integration with telephone services can contribute to two objectives:

1. Preventing the user from having to enter too much information.
2. In the purchasing process of your product or service, personal contact could be necessary to persuade the visitor to buy.

The need for entering information will become an unavoidable obstacle for many users. Take, for example, creating a user name and a password to place an order via a mobile phone. To avoid requiring the user to sign off completely to create an account via the PC, the user can directly contact the provider's telephone help desk, where someone can enter the user's information for them. Next, the user can use the service with his or her own account and he or she will have the account information sent via SMS or email. Another example is the entry of a personal profile with an online real estate agent. The visitor has indicated that he or she is interested by visiting this site. The opportunity to discuss preferences with the realtor gives the user much more added value than simply filling in a form. The final possibility is a telephone number for questions on how to use the service. Where a Web site has space for frequently asked questions and a help function, this would not be as easy for a WAP site. Sending a short email can be done, but is not always welcome. For the user it is easier to call the help desk with questions on the service. This prevents the user from being disappointed and signing off.

Personal contact is often the determining factor in the purchasing process because the product or service offered demands this or because your company's added value includes personal advice. Integration of the WAP site in your selling process is valuable in these cases. Take, for example, a software vendor's site. As a visitor to this site you find an unfamiliar software package with new possibilities that are useful to you. You want to know more about this package before you decide to purchase it. With two clicks you call the help desk, ask your questions, and purchase the software. Giving the customer the choice between a personal conversation and an electronic transaction is advisable for many sites. The value of personal contact in the purchasing process should not be underestimated! Some information is more easily absorbed visually than aurally. Think in this case of account numbers, amounts, telephone numbers, product codes, or statistics. For extra information or questions on scheduling appointments, verbal communication is preferred. Integration of speech services in your WAP service depends on the objective of your WAP site and the importance your target market attributes to personal contact. If your company already has a help desk or in-house sales staff, these services can be used for WAP as well. The additional investment cost for the backbone of the organization is minimal in this case; the relatively high exploitation costs of a telephone help desk are already taken into account without mobile service.

Availability

High availability makes mobile telephony a unique medium. In comparison to the few hours the average Internet user is online every day, the 24/7 availability of the mobile telephone is a marketer's dream. The availability of Internet access in different locations than the user's study provides many opportunities. Take, for example, the possibility of interaction with television. There are many referrals to telephone numbers and Web sites in television commercials. The effectiveness is much higher if the site can be viewed instantly or can be stored in the telephone with the click of a button to be viewed later.

TV shows also present telephone numbers or bank account numbers for charity events. If action can be taken right away, effectiveness is much higher. It often happens that during a discussion you have a question that requires an immediate answer. It would be easy to just look for the information right away without having to get up. Availability at work creates opportunities for other applications. Sometimes you look for the telephone book or the yellow pages. A computer with Internet access is not always at hand. Company-specific services probably offer the most added value in the workplace. Other examples are workplaces where the PC with Internet access is barely available, such as restaurants and retail outlets. The mobile telephone can be used to fulfill tasks that are currently fulfilled by a fax or a regular phone, requiring much more effort. Tasks such as placing orders, paying bills, checking for delivery times, or checking order status can be done faster with mobile telephony. Some shippers already offer tracking and tracing services via WAP.

The most obvious moment for the use of mobile telephone is on the road (not meaning usage while driving). When we take into account the time we are on the road, visiting friends, family, or on business, there are great opportunities for existing services and for new services. Existing Web sites with services such as maps, public transportation information, and traffic reports can generate much more revenue than previously. Voice-enabled applications can offer the same services while driving a car. Existing services can be expanded. Directory providers cannot only offer address information this way, but also a map together with directions. Completely new services will appear that go beyond what we can imagine. Think of a walk through the city center with directions and tourist information via the mobile phone, or checking the ingredients while you shop to find out what you really need before you buy the product.

Barrier to Purchase

The low barriers to purchasing a mobile phone with WAP make the Internet accessible to people that will never use a PC, a very large group of people from several social groups. The suppliers that want to make these people their customers will have to work with a different frame of reference. The previously mentioned interaction of WAP with television shows will supply these target markets with many new applications.

User Threshold

From the perspective of the employer, it is worthwhile to supply traveling employees with a mobile phone. The account manager, the consultant, and other mobile professionals already have a mobile telephone in most cases. For them WAP is a logical addition to the options. Employees who can be reached with mobile Internet services include mechanics, drivers, or construction workers. Many of these people have mobile phones, but no PC. Several manufacturers have developed mobile phones that can be used outside because of waterproof and shockproof membranes.

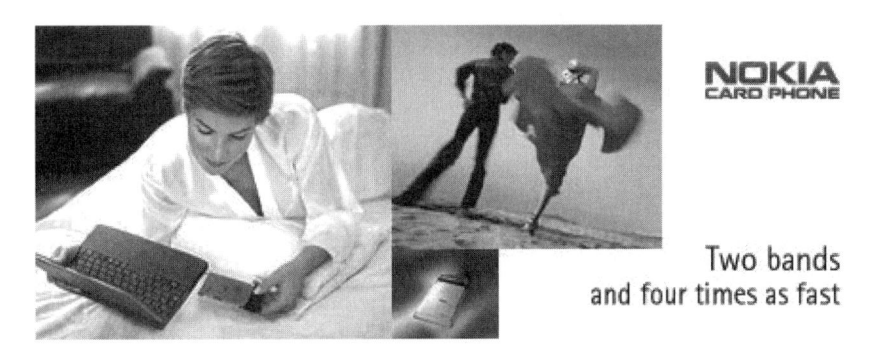

Figure 2.7
Nokia Cardphone 2.0 for mobile laptop use.

The mobile professional often has a laptop and a mobile phone (see Figure 2.7). However, the user threshold is too high to access the Web whenever you want. It takes up valuable time to unpack the laptop, log on, and connect to the intranet. For the calculation of mortgages, this is not a problem. For checking the news or a calendar, this is too much work. The mobile phone offers instant services, so users can look at WAP sites easier, faster, and therefore more frequently than Web sites. The sessions will be shorter than those on the Internet. Users can check the traffic reports before they start the car or check the news while waiting for the bus.

Personal Use

Reaching customers in a direct and personal manner is a marketer's dream. The possibilities seem endless, as long as they are used in the appropriate way. The user determines the appropriate way. Portals such as Yahoo! and CNN offer tailored information, and many other service providers use a personalized environment for financial transactions. With a WAP site, a personal approach contributes to a faster and easier use of the service. When more information on the consumer is retained, the consumer will have to enter less information. The main advantage of mobile telephony is that the service provider can recognize the user. The relationship is stronger than a cookie via the Internet, as the customer always uses the same SIM card (Figure 2.8) with the same number. On the Internet, a site can be visited from different PCs by the same customer, without the site being able to determine that it concerns the same customer. Because of this, logon procedures are no longer a condition for most personal service providers. It remains to be seen if the logon procedure for mobile telephony

Figure 2.8
SIM, Subscriber Identity Module.

will disappear. It gives the user a sense of security to provide authorization prior to making a payment. The risk of unauthorized use in case of loss or theft of the mobile telephone cannot be underestimated.

With a WAP site, the user selects what he or she wants to watch (pull). As soon as the pull of the WAP site is combined with a push medium as SMS, another situation arises. SMS is the personal mobile equivalent of email to a certain extent. Email accounts are often shared, people often have more than one account, and they check their email accounts at different frequencies. At the moment a user receives an email, he or she will often not notice this directly. A mobile user receives an SMS right away, assuming that the phone is switched on. In this sense, comparison to ICQ seems more relevant.

Unwanted SMS messages on the mobile phone will annoy the user more than unwanted email. In this case, permission marketing has to be activated to supply services that add value for a user. An example is a realtor's site. On the Web site you sign on and indicate what kind of house you want, where you want it to be, and the price you are willing to pay. You also indicate that you want to receive a mobile message as soon as a house that meets your criteria becomes available, so that you can react right away. The realtor groups the houses that meet the criteria and posts these on a limited-access section of the Web site and the WAP site. The customer receives a message via SMS if a house is available. The customer then can read the information via the PC or mobile phone and take action.

Security

The big advantages of the supply of services via a mobile telephone are the increased security of transactions and simpler and faster security mechanisms for the customer. Paying with a couple of clicks is much simpler for the customer than current methods of payment. Installations of payment mechanisms at the suppliers are no longer necessary, so that the cost per transaction is low. Mobile transaction possibilities mean that transactions can be made wherever there is GSM coverage, without a need for cash. A mobile phone fulfills the function of a payment mechanism. Paying for products at the market will be done electronically in the future and business transactions can be finalized instantly.

Invoicing

Fulfilling small payments for services via the mobile telephone has been around for a long time. The higher priced service numbers are an example of this. KPN Mobile's @info has offered the ability to charge amounts up to $1 via the mobile invoice since its introduction. For larger amounts, regular methods of payment such as credit cards and bank payments can be used. Existing customers can also order on account. For an entrepreneur, this means that there are various possibilities and even better possibilities are being developed for easy payments via the mobile phone at lower costs. Administrative departments can realize cost savings by making payments more efficient. Shorter terms of payment contribute to a lower level of tied-up capital. Customers can receive a mobile reminder of when payments are due. This lowers the costs associated with repossessions.

Price Perception and Pricing

The much-heard comment that almost everything on the Internet is almost free is not true. Valuable information is often limited to subscribers only and Amazon.com asks for money for its books and CDs. In the matter of price awareness of mobile telephony, the opposite is the case. The classic initial creaming strategy followed by increased competition and differentiation has resulted in large price decreases for devices, subscriptions, and call minutes. The user knows that mobile services cost money and accepts this. Upon the introduction of @info, the complaint was that the prices were too high. This was partly caused by the speed of the service. Even if the speed is not fast, the cost can be high if it takes a long time to find the information. The rates of a number of service providers are regarded as high. The service providers are free in setting their prices, but have to experiment with pricing policies for this new medium. It is clear that information has to be up-to-date. The comparison with other free news sources such as teletext and the radio is regularly made. Separate from the perceived value of a product or service, a fee could be an important contribution to the business case of many suppliers.

Large numbers of users could yield a substantial amount, even if the amount per user is limited. Finally, the price perception with mobile telephony offers the possibility to ask a price for a package of services, for example, WAP, Web, and help-desk support.

Similar to the Internet, price comparisons will become popular. The influence of price comparisons via mobile telephone to store prices will be much larger. There will be services that can tell a user right away if the price on a product is the lowest price, or if the same product can be purchased somewhere else for less. The transparency of the prices will increase, resulting in a leveling of prices that will differ for each product category. During the weekly grocery shopping trip it is uncommon for a customer to check if the product is cheaper at a different store. For goods such as clothing or electronics, this is not the case. In this case, the savings can be much larger per item.

Integration Possibilities

A service provider has to wonder if the service it offers via the Web could be enriched with mobile access. A supplier of services via the Internet can reach customers better and more often via the mobile phone. This can be done by making existing Web sites accessible for the mobile phone. Think of reading email or bank details. Often, Internet and WAP can both be used for an integrated service or WAP can be used for providing services that were not possible in the past. Examples are the following:

- Services for mobile use that can be set up via the Internet
- Personal services accessible via WAP and PC, for which the PC is used to enter the personal preferences
- Combined services for which orders are taken via the Internet and payments are made via mobile phone
- Services for use via the Internet and warnings via mobile phone

It is important to combine the strong features of mobile telephony with the Internet and tune them to the profile of your target market and the nature of your service.

Table 2.2 Overview

Hardware characteristics	The mobile telephone has limited functionality in comparison with the PC. Services have to be optimized for the available functionality. Services have to support several device configurations. Ease of use is crucial with regard to the size of the screen and the limited input possibilities.
Speed	The limited data speed requires information that is adapted to the medium.
Up-to-date	The accuracy of the information determines the frequency of visits and the price level. For many information services, up-to-date information is a must.
Relation with telephony services	Integration with telephone services circumvents the limited input possibilities and offers the opportunity to supply more information.
Availability	High availability both inside and outside the company allows for many new applications, but also creates the need for current information and personalization.

Table 2.2 Overview (Continued)

Threshold to purchase	The low threshold to purchase a product results in a large target audience. Mobile Internet users have other needs compared with the average Internet user.
User threshold	Low user threshold results in instant use. Up-to-date information and speed of use are factors for success.
Personal use of connection and services	Opportunities to reach customers directly provide new opportunities and demand personalization.
Security	Transactions via mobile telephone offer opportunities for cost savings and higher revenue.
Billing possibilities	Various possibilities, as the relationship with the telecom provider or banks can be used for invoicing purposes. This lowers transaction costs.
Price awareness and price determination	Opportunity to ask fees for the services from the end users to generate revenue via WAP. Price determination depends on the nature of the product, availability, and the price of mobile initiatives.
Integration possibilities	Integration with Internet services could supply the customer with added value in a number of ways.

WAP VERSUS THE INTERNET FOR THE SERVICE DEVELOPER

For the development of WAP, the Internet architecture and the Internet Protocol have been taken as a starting point. Originally WAP stems from the mecca of the Internet world: Silicon Valley. Unwired Planet (now a public company and renamed Openwave) is one of the initiators of WAP (see the section devoted to Openwave in Chapter 1). For the continuing development of WAP, Internet developments are closely monitored. WML is a markup language based on tags, specified as an XML document type. Existing XML authorization tools and HTML development tools can be used for WML without any problems. From the WAP gateway, HTTP 1.1 is used for the Internet so that existing Web servers and techniques can be used for WAP. For the WAP service, a separate URL can be used on the existing Web address. It then has to be checked if HTML or WML content has to be supplied, depending on the type of browser. From the former it can be concluded that the use and the applications of

WAP services differ on a number of aspects from the Internet. There are also a number of differences in the development of the services.

Characteristics of a Mobile Phone

The small processor capacity of the mobile device limits the designer, especially in combination with minimization of the energy consumption in favor of standby time. Offering services for which much information has to be entered via the mobile phone is not advisable, but in most cases it is necessary to enter some information.

In the design, solutions can be developed to minimize data entry. If the information can be displayed in a limited number of options, the user can scroll through the options and select one. With larger amounts of text, limited data entry will suffice. For example, the first four letters of a city or the first four digits of a zip code. To place the relevant information on a small screen, much of the information available on the Web site has to be omitted. For example, a rich text document will have to be stripped from many graphics and annotations to be used successfully via the mobile phone. Text information will have to be abbreviated. More and more applications have become available that support the automatic conversion of HTML to XML and create semiautomatic context-dependent summaries. The possibilities for user support with help functions and manuals are limited. There might be an opportunity for a WAP wizard, an intelligent agent that helps with the use of certain services.

The large variety in screen sizes, microbrowsers, and keyboard functionality with mobile telephones provides the developer with a challenge. Developing for the minimum available functionality is a possibility and could be interesting for simple text-based applications. For more complex applications with graphics possibilities, it is advisable to take the screen size into account. Many people specifically purchase a telephone with a large screen and they will not be satisfied with graphical services that are based on the minimum screen size. If the service developer does not use this extra space, the service is falling short. Fortunately, the mobile phone indicates the device and browser identification with every message, including the browser type and the software version. This way the most common devices can be optimized, and for the other devices a standard version can be used. With the introduction of new WAP versions, the matter of backward compatibility is important as well. Competition between browser and device manufacturers will always exist. Sites that are aimed at the U.S. public may have to support advanced two-way paging systems and Microsoft and Palm PDAs. In conclusion, with regard to development, we need to consider the large flexibility of different appliances, as the target audience for mobile services is too limited otherwise. This diversity will only grow with the rising penetration of all kinds of PDA devices.

Speed

Speed of the network determines the amount of data that can be sent. The speed of mobile networks will increase greatly in the next few years, but will remain lower than the speed of regular networks. The way mobile phones are used, egg timers on the screen are unacceptable for most users. This means that the designer has to give priority to speed and ease of use. Right now speed is more important than appearance. Of course, games will be developed with color graphics. The use of graphics in applications will depend on the type of service. By developing menus, offering intuitive navigation, and striving for uniformity in the use of functions and naming, the designer can positively influence the perception of speed.

Integration with Speech

The importance of the integration of speech telephony in a WAP service was already mentioned in the previous section. Already presented telephone numbers can be dialed. The possibilities for further integration increase with the implementation of the Wireless Telephony Application (WTA) in the WAP 1.2 release. The expectation is that operators will integrate WTA in their networks in a secure environment. They may not want to provide access to the telephone infrastructure for third parties. This means that the operators will give developers an interface description to build services that use WTA features.

The high availability has the same consequences for the service developer as the Internet. Help desks may have to extend their hours. Management of services has to be planned carefully in off-hours. In most cases, somebody has to be around or on standby to solve technical problems. The combination of WAP with television shows makes capacity planning for both the number of dial-up lines and the connection to the server of the service provider necessary. When the first 1-900 number was used to cast votes on a popular television show, the capacity was often insufficient to receive all calls. The dial-up capacity is not a problem in the case of WAP, as the impact is distributed among all parties with dial-in facilities. The number of dial-up facilities will increase rapidly; this can be compared to the strong growth of the number of Internet service providers. For the designer, the capacity of the server has to be capable of handling an important event where people call to cast their votes.

Purchase Threshold

The low price threshold for purchasing a mobile phone will lead to groups of users that are attracted to the use of WAP services without experience with the Internet. Therefore, these users do not have the same frame of reference as the average Internet user. It is expected that the WAP user group will mostly exist of experienced Internet visitors in the beginning, but later a considerable number of users, barely familiar with the Internet via the PC, will join. The type of user has an impact on the design of services.

Services aimed at the target audience without experience cannot use the Internet as a frame of reference. For this group, ATMs, television, and teletext may provide a frame of reference. The first WAP services can be compared to teletext in that they consist mainly of text. Control via a remote with a limited number of buttons can be useful. The short function labeling, hyperlinks, and the way soft keys are treated will provide problems for many users. Moreover, the possibilities for help functions are limited in comparison to the Internet. Many users will have little understanding of the familiar error codes and difficult key combinations. A certain uniformity of services could help in this case. For the designer, the characteristics of the target market have to be clear to create a good service design.

User Threshold

The lower user threshold leads to instant use. The frequency and short-term use of WAP services will increase, especially when the dial-up time is nullified with the introduction of packed-switched networks. In combination with high availability, people will also be able to use services all the time. Frequent and short-term use demands that the services are up-to-date and fast. For service developers, this means that the information must be created and maintained quickly and easily. Besides availability of people with these capabilities, system support will also have to be developed. As the information has to be suitable for different types of devices, a distinction between content and format is necessary. Existing content management systems developed for the Internet are the first step in the right direction. A number of these systems have been expanded with functionality suitable for WAP and the Internet.

Users want to find information fast via their mobile phones, and they want to reach their goals in a limited number of steps. The designer will have to find a balance between the number of layers a service has and the need for navigation options and recognition points for the user. For the design of the @info portal, we chose to use simple graphics for the various topics. Even though this makes the service one layer deeper, it helps the user navigate through the service. Using the graphic, the user knows where he or she is. Users can also be encouraged to bookmark a service in their portals or their phones so they can access the bookmarked services faster, similar to the Web.

Personal Use

Personalizing the services increases the speed of use. This means that users receive information targeted to their needs, preferences, or behavior. Approaching the user personally means that the wishes and preferences of the user have to be monitored closely. The stored data has to be available online, preferably combined with the behavior and the prediction of the expected behavior. On the Internet, there are plenty of examples of services that keep track of their customers' preferences. Amazon.com is one of course, but there are many others. What makes WAP special is that the user expects this.

Because it is his or her own mobile phone, the average user expects that the services are also personal. The user that dials into voice mail will hear his or her own messages. Based on customer identification, every service developer can recognize frequent visitors and target its services. The question is whether mobile Internet service providers give out the mobile number as a customer identification number. As many users give their mobile numbers to a limited group of people, the unrestricted use and sales of the mobile numbers to third parties will meet opposition from some users. For that reason, KPN Mobile does not give out GSM numbers as customer identification numbers, but instead uses unique customer identities. Users can then be recognized by the service provider, but cannot be reached via the telephone. The personal approach of customers puts demands on the design of a customer database, the necessity of linking to other systems, and the possibility to make individual customer preferences available online.

Security

The possibilities to execute transactions can have a large impact on the design of the service. The WAP standard, in its current form, offers a protocol for transactions. This is the Wireless Transport Layer Security (WTLS) protocol, based on the standard Transport Layer Security (TLS) protocol (previously called Secure Sockets Layer or SSL). WTLS provides data integrity, privacy, authentication, and security against denial of service. The WAP gateway communicates transparently and automatically with Web applications that use TLS facilities. In the next few years, the security for executing transactions will increase. In Chapter 3, we pay attention to these developments. It is no longer a problem with the available technology to execute payments with a credit card. The user has to enter the credit card number. In several locations, WAP pilots are conducted with mobile phones with two slots: one slot for the SIM card and one extra slot for the credit card. The user can then enter the credit card into the phone to pay. Service providers can supply the existing customers with a user name and access code to allow for orders on account.

Billing Possibilities

KPN Mobile offers its customers a safe and simple method for micropayments via the mobile invoice. KPN Mobile has realized a link between the service and the SIM card of the user. In the case of day tickets, a user can access a specific service for a certain fee for 24 hours. Accessing CNN top stories and news for a day costs $10. With day tickets, KPN Mobile takes over the transaction processes. The service provider will not have to execute any additional activities. The service provider can also execute payment on a transaction basis. This has the advantage that the provider can charge different prices for different aspects of the service. For this the service provider needs to generate a call detail record (CDR) per transaction with the unique customer identification and the amount and supply them to KPN Mobile. The CDR is used by KPN Mobile to process the transaction on the user's invoice, in which the unique customer identification is translated to a mobile telephone number.

In Japan, NTT DoCoMo offers its service providers similar invoicing options. In their case a monthly fee is charged, for example, a subscription to a cartoon character as a screensaver that changes daily. Billing via the mobile invoice is expanded further with the introduction of WIM (Wireless Identity Module). The link of WAP and WIM integrates the SIM card in WAP, with which complete end-to-end security is guaranteed.

Integration Possibilities

Integration with the Internet is important in a number of ways. As described, existing Internet service can also be made available to mobile telephony. The way in which a Web site is constructed determines the degree of difficulty. Services with the objective of displaying information are easy to produce. It is important to keep the same look and feel and uniformity for the different interfaces. Examples are the consequent use of terminology, identical site structure, and (if necessary for WAP) uniform logon procedures. As long as Web and WAP sites make use of the same databases, integrated services are possible. The credit card payment mechanisms can also be used for WAP, just like email confirmation, logistic processes, and help-desk support. For a designer who has to create a WAP service in combination with an existing Web site, it is more or less the same. The condition is that the information output is stored separately. To offer new WAP services, the support functions also have to be created. The implications depend on the service and the existing infrastructure. Reserving a parking space in a parking garage seems like a great application, but in reality it is difficult. How can the user park the car in a reserved spot if there is a line of cars waiting at the entrance? Will every parking garage have a separate entrance for drivers with a reservation?

A big difference between WAP and the Internet is the attitude of the user. An Internet user sits on a chair and watches the screen. Services on mobile phones are used for a short period of time at various places. Often, users will briefly look away from the screen before crossing a street. It is therefore important to estimate the user situation before the design is created or to try the service in different situations. The user situation could result in adding a pause button for a game or an extended input time for a shopping service. Also, the use of flash screens (text that disappears after a few seconds) has to be considered. Certain WAP services in the car are examples of this. Even though directions and traffic reports are very relevant, a motorist is not in a position to use this information in a safe manner.

Table 2.3 Overview

Hardware characteristics	Impact of the demand that services have to support different types of devices. Services have to limit the use of processor capacity. A user interface has to minimize the input of data and take over the help functions.
Speed	Slower speed forces efficient use of the bandwidth and the information and navigation tailored to the medium and the target group (and higher).
Relation with telephone services	Integration with telephone services is part of the design. WAP 1.2 standardizes the interfaces with WTA specifications.
Availability	High availability creates a need for personalization and additional features.
Threshold to purchase	Groups of users do not only have the PC and Internet as a frame of reference. This places demands on the design, especially the ease of use.
User threshold	An increased need for current information and speed of use have consequences for content creation and design.
Personal use of connection and services	Possibility to reach customers directly places demands on database management.
Security	Transactions via the mobile telephone create the need for links to other systems.
Billing possibilities	Various possibilities. Besides the existing possibilities of the Internet, the relation with the telecom supplier or bank can be used for billing purposes.

Table 2.3 Overview (Continued)

Integration possibilities	Integration with Internet services creates additional demands on the design. Uniformity is the key.
Applications	Targeted search by customers, multitasking, etc., places demands on the formulation of the services.

WAP AND CORPORATE NETWORKS

What goes for the Internet generally also goes for intranets and extranets. Employees of companies want access to planners, email, and intranet without the limitations of desktops (see Figure 2.9). The mobile employee especially values access to relevant information via the mobile phone. People find it convenient to not always have to contact people in the office for information. Direct access to company-specific files and services such as up-to-date customer files, the ordering application, stock application, or the internal telephone list increases the effectiveness.

The experience of KPN Mobile and other operators such as Norwegian Telenor indicates that WAP is only a part of the puzzle. The use of corporate databases requires more than the purchase of a few WAP telephones and a couple of hours building a WAP site. Presenting information from self-built applications on mobile phones, the organization of security, setting up procedures for access permission, and the setup of system administration are activities that are important in creating mobile access to company networks. In general, the developers of the intranet and extranet applications are best equipped to contribute to the connection of WAP telephones because of their experience with the databases, their knowledge of the organizations, and the project-oriented approach of their employees. Companies such as Atos, Origin, Cap Gemini, IBM, and numerous others have founded knowledge centers and are working on solutions with telecom operators and customers.

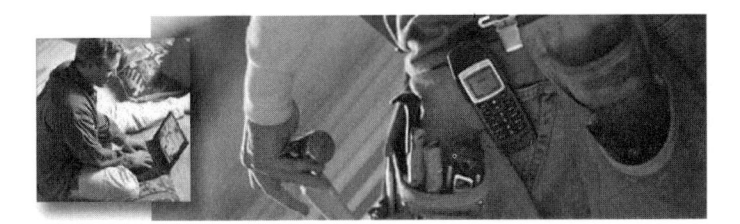

Figure 2.9
Access to the intranet wherever you are.

The results are promising when we look at the first findings. They indicate that the services are easy to use, more flexible than existing solutions, and highly customizable. A large advantage is that these users are already familiar with the applications in the desktop environment. On the other hand, there is room for improvement. As the menus are not yet user friendly enough, the services need to be better adapted to the smaller format and the way the employee absorbs the information.

The mobile telephone is mainly used for short messages, often complementary to the laptop. The most used functions are reading and deleting, so that the entry of too much data can be avoided. Even though the developers appreciate the technology, they are skeptical about the security of corporate access and speed.

Security is an issue that cannot be left unnoticed. Many companies are very careful when it comes to linking company networks to the Internet by using firewalls, dial-up procedures, and certificates. For mobile Internet, the same complaints can be raised.

Information on the GSM radio waves is sent encrypted and conforms to the methodologies that GSM speech has included in the standards. WAP uses WTLS, and from the WAP gateway, contact can be made with the company network outside of the Internet, via ISDN. For many companies that want to offer WAP intranet access to their employees, an important question is: "Do I have to purchase a WAP gateway or do I let employees use the mobile operator's gateway and do I connect my intranet in a secure way to the WAP gateway of the mobile operator?" A great benefit of the use of the WAP gateway of the operator is that the care for sufficient dial-up and user capabilities, keeping the gateway up and running, upgrading to the next WAP versions, the right setups in the devices, and more of such technical matters lies with the operator. The company only has to worry about the development and maintenance of the WAP applications for its employees, or it could even outsource it. An advantage of the use of an owned WAP gateway is that the company has more control over which services the user can use besides the company's applications and the security measures taken. With the operator, the user often has access to all services, whereas a company often wants to place restrictions on use. Banks will almost always choose their own gateway for the development of applications for mobile banking. In the WAP Forum, people are working on possibilities for gateway routing. This means that a customer can access WAP via its normal WAP service and that the connection is rerouted from the gateway previously used to the bank's gateway at the moment the bank service is selected. This way the bank's higher security measures can be met.

A logon procedure will be necessary for access to the company network to limit the risk of unauthorized use by theft or loss. There will also be differences between the employees of a company. A mechanic needs access to inventory systems and planning programs. Sales figures by region can be kept hidden from the mechanic, but not from the sales manager. This demands that existing software packages be expanded to allocate these types of access for mobile devices as well.

Widely known standards such as Lotus Notes, Microsoft Exchange, and Microsoft Outlook have mobile access. The members of KPN Mobile's @info service have been able to use Lotus Notes since December 1999. A number of Microsoft developers are working on WAP applications. IBM makes source code available for synchronizing data from the regular and mobile environment. Expansion of middleware for e-trade and wireless transactions is under construction. IBM has also launched a product, Websphere Transcoding Publisher, that enables automatic translation of information to mobile formats. Siebel's eBusiness product offers an Internet connection for both regular and mobile appliances. This is merely a selection of the many products and services developed by software suppliers and solution providers to make existing information and applications available as easily as possible.

3 The Rapid Development of Mobile Communications

In this chapter...

- Mobile Communication   82
- Different Roles in Mobile Communication   86
- Mobile Phone Usage   96
- Mobile Company Networks   98

Mobile communication is a very fast growing market. In 1992, there were 23 million mobile phone users worldwide (still analog of course). It's expected that the number of users will reach 1 billion in late 2002 or 2003 (see Figure 3.1). It took 130 years to reach the same number of fixed lines.

The market for fixed Internet use has had a similar rapid development, but the growth started later than the growth in mobile communication. In both the mobile communications and Internet industry, developments are taking place at a very rapid pace. What takes certain industries years to accomplish happens in months in these industries.

In the first two chapters, the functional and technical aspects of the mobile Internet were discussed. Just as important are the developments of the value chain, the business models, and the most important players and alliances. This chapter is the first of three that will cover these developments. This chapter describes the developments in the market for mobile communication, the players, and their parts. Special attention is given to the 3G auctions, the concentration wave among mobile operators, and the rise of Short Message Services (SMS) as the mobile equivalent of email. Chapter 4 gives a short summary of the roles, the use, and the development of the "traditional" Internet. Readers who are familiar with the Internet may skip Chapter 4. Chapter 5 is dedicated to the mobile Internet. The roles in the mobile Internet value chain, the conflicts among these roles, and the critical success factors are also covered in this chapter. Close attention is also paid to the market of hybrid mobile devices, the importance of banks with respect to payment via mobile phones, and the development of mobile Internet outside Europe. Developments in Japan are described because the Japanese are using the mobile Internet on a very large scale already.

MOBILE COMMUNICATION

In early 2000, there were more than 500 million mobile connections worldwide. Even though most early adopting countries started with analog networks and later switched to digital networks, about 90 percent of all connections are digital. The most widely used digital technology is GSM. More than 69 percent of all the mobile connections worldwide are connections to a GSM network, mostly in Europe. GSM first became the standard in the European Union and later also in other parts of the world. The largest advantage of this standard is the fact that among mobile operators in different countries so-called "roaming" agreements can be made. The mobile phone can thereby be used on GSM networks in different countries. More details about GSM and other mobile standards like PDC and CDMA can be found in Chapter 1. In the United States, 62 percent of the customers use a digital service. Different standards in many parts of the United States and Europe make it difficult for European travelers to use their mobile phones in the United States and vice versa. Often a business traveler therefore carries different phones with different numbers, using different standards.

Mobile Communication

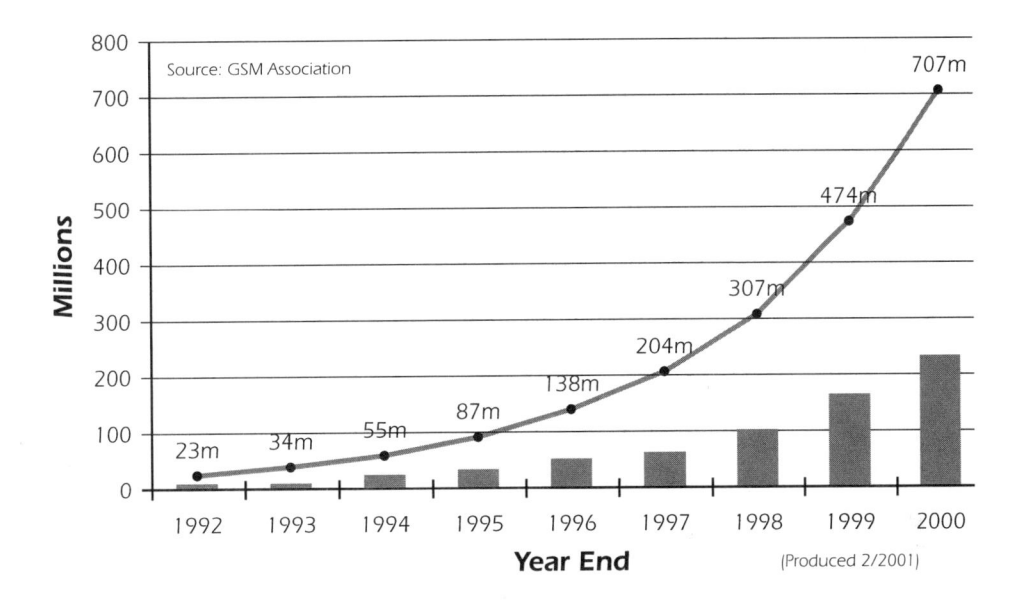

Figure 3.1
The worldwide growth in cellular subscribers (bars = in-year net gain).

The territories of Europe and Asia Pacific take the lead in the number of mobile phone users (see Figures 3.2 and 3.3). In Africa, the Middle East, and Latin America, the numbers are much smaller. The United States and Canada are in the middle on the scale of users with 35 percent penetration in the United States. Within each territory,

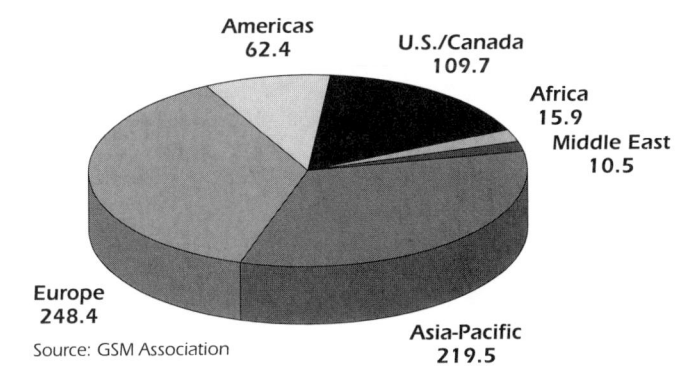

Figure 3.2
The market share of cellular subscribers by world region in millions (December 2000).

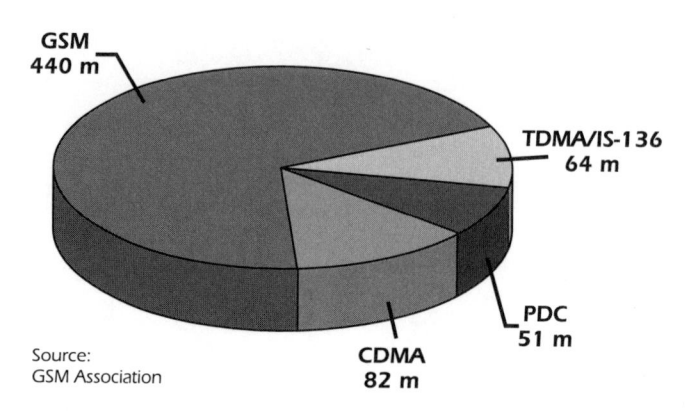

Figure 3.3
Number of subscribers per network technique (December 2000).

the differences in penetration are large. The market penetration is defined as the number of mobile telephone users divided by the number of people (per capita). In Europe, Finland takes the lead with a penetration of 73 percent as of September 2000. Finland is followed by Sweden, Norway, Italy, and Portugal, with penetrations of more than 67 percent. The UK, the Netherlands, Spain, and Switzerland are in the middle position, with penetrations of around 60 percent. Germany, France, and Ireland are Europe's laggards, with penetrations between 35 and 50 percent. Growth in numbers of users will take place in Asia, followed by less penetrated countries in Africa and the Americas. In Europe, the focus is on increasing usage by offering additional services to mobile customers. Offering differentiating services is also important for mobile operators to prevent users from changing networks in these competitive markets.

In Europe, the European Community played an important role in the development of the mobile telecommunications market, first by standardizing GSM within ETSI and then by adjusting regulations. On recommendation of the European Commission, national governments allowed competition in the market for telecommunications and the state-owned national telecommunication companies were privatized. The privatization has become effective in most countries, but at various intervals. In England, British Telecom's stock was traded in 1984. In the Netherlands, PTT went public in 1989, and in 1994, its stock was traded under the name KPN. In 1998, the postal and telecom divisions were separated and stocks were traded under two separate companies. In France, France Telecom was privatized in 1996 and its stock was first traded in 1997.

In some European countries, the first GSM network was built by the monopolistic state-owned telecommunications company. In other countries, two licenses were granted from the start, for the GSM network or its predecessor.

UMTS Auctions

In Chapter 1, we covered UMTS as the latest technology that enables multimedia for mobile phones (see Figure 3.4). UMTS uses radio frequencies other than those used by the current 2G networks. In the introduction of UMTS, the European Commission again played an important part. The European Commission decided that all licenses for UMTS frequencies for all European nations have to be distributed by 2001 at the latest. The national governments are free to decide how they want to assign frequencies to the telecommunications companies. Governments make frequency space available to companies for a period of, for example, 20 years. For GSM networks, these frequency spaces were often given to the incumbent operator for free. The privatization of the European telecommunications sector gave other parties the chance to compete with the incumbent. The second mobile operator was often selected based on a "beauty contest" without having to pay for the license. Later national governments realized that there was money to be made by auctioning licenses for mobile telephony, which several did for GSM licenses.

In Europe, most UMTS licenses were awarded in 2000. Two methods were used: the auction and the beauty contest. Countries like the United Kingdom, Germany, Italy, the Netherlands, Switzerland, and Austria used the auction method with very different results. The total revenues from the auction were 50.5 billion Euros (about $48 billion) in Germany, 37.5 billion Euros in the United Kingdom, but only 136 million Euros in Switzerland. The beauty contest (selection on business plans) can also be a method for governments to earn money by selling UMTS licenses. The French government asked 4.95 billion Euros per license, whereas the Finnish government gave them away for free. The world's biggest mobile operator, Vodafone, had

Figure 3.4
Prototype of a Nokia UMTS terminal.

obtained UMTS licenses in Sweden, Germany, the Netherlands, Portugal, and the United Kingdom by mid-2001.

The European Commission would like to see the revenue of the auctions and beauty contests reinvested in the telecommunications sector. The politicians, however, see this money as additional revenue and are debating the allocation. The European parliament has criticized this auction, as telecom companies from countries without license auctions have more financial freedom to bid at auctions in other countries. The parliament has also stated that the high prices associated with the licenses will negatively affect the innovation of services. The telecommunications companies have voiced their opinions with regard to the artificially created scarcity and speak of hidden taxation. In their eyes, the frequency space is not scarce, as there are plenty of frequencies available for satellite communication. In the end, the bidding parties themselves determine the amount they are willing to bid, but a UMTS license seems to be a requirement for operators to nullify the expected frequency scarcity of the current network, and to become successful in the mobile Internet market. To lower total investment cost, mobile operators try to share the same UMTS infrastructure. This is under discussion with the various national regulatory bodies, who fear that sharing infrastructure will decrease competition.

In Japan, three 3G licenses were issued to the existing mobile operators NTT DoCoMo, Japan Telecom, and KDD + IDO. In other Asian countries and in the United States, the 3G licenses will be awarded in 2001 or 2002 via auction or beauty contest.

DIFFERENT ROLES IN MOBILE COMMUNICATION....

In the mobile telecommunications market there are several roles. The most important roles are the mobile operator, the service provider, the network equipment supplier, and the device supplier. With the addition of value-added services such as WAP and SIM toolkit applications, roles are added, but we limit ourselves to discussion of these four roles here.

Mobile Operators

The role of mobile operators is construction and exploitation of the mobile networks. A mobile operator can compete on different levels: price, quality, and service. Price competition has been the most widely used. In many countries the entry into the market of more operators often means a steep decrease in rates for mobile telephony. This decrease in rates was often covered up with a simultaneous introduction of a different method for determining rates. The first operator often charged by the minute and asked a standard fee per conversation. This made the published rates look lower than they really were. The new competition put an end to these monopolistic practices. In the Netherlands, Libertel was the first to introduce paying per second, copied quickly by KPN Mobile. New rate frameworks such as the introduction of bulk calling and discount rates to specific destination numbers were also offered. With bulk calling, the caller buys a quantity of minutes at a low rate. In the United Kingdom, the Orange operator allows you to call at discounted rates if the person you are calling also has an Orange connection. With this, operators are trying to start the flow of friends signing up friends.

Another form of decreasing rates is the subsidizing of mobile devices. The mobile operator pays for the device in part when the customer purchases a subscription to the operator's services. In the United Kingdom, France, and Germany, this has now become common practice. In many countries like Belgium, Finland, and Italy, this practice of decreasing handset prices is not permitted, as the revenue from the network is used to subsidize the purchase of a mobile device. In some countries, this is regarded as cross-subsidizing and is thus illegal.

Some rate structures are not as easy to imitate. Italian operator Omnitel was the first to introduce prepaid phones successfully. Instead of a monthly subscription with an invoice at the end of the month, this allows users to purchase minutes before they call. Even though the rates are higher, prepaid is very successful in many countries, because the caller only pays for calls made. No fees and prepaying give users more control over their spending. The anonymity offered as a result of prepaid is also valued greatly. Prepaid has now been introduced in most countries. Worldwide, 40 percent of the new mobile connections were prepaid in 1999. Some operators believe 75 percent of their customer base will eventually be prepaid users.

Another way of competing is the competition based on quality. The quality factors change as the mobile telephony market matures. During the startup phase of mobile networks, coverage is the most important quality parameter. In addition to coverage, other quality parameters such as call completion (the number of calls made without interruption) and quality of sound are also important. An example of differentiation of quality was demonstrated during the launch of third and fourth operators in many countries. They used GSM networks based on 1800 frequency instead of GSM 900 frequencies and introduced EFR (Enhanced Full Rate) to claim superior quality. Besides the quality of the network, supplementary services are also competing factors. Examples include voice mail, SMS, and, of course, WAP services.

The last competing factor is service. It is difficult to distinguish products based on service, and only a few operators compete this way. In a rapidly growing market, it is difficult to compete based on service. Most users who obtain a device and connection do this for the first time. A nice device, price, and quality are more important factors for these people than service. As soon as the market is saturated, however, competition based on service intensifies.

Brand names also become more important when mobile telephony becomes a commodity. In the United Kingdom, Orange has created its own brand in an unconventional way. The brand is positioned offering freedom, choices, independence, and always staying updated. From the start, Orange has distinguished itself from Vodafone and Cellnet who have placed more emphasis on technology. The Orange brand has increased the value of the company, as its customer loyalty is larger than that of the competition.

Concentration Wave in the Telecommunications Industry

The privatization of the national telephony markets has resulted in a substantial growth of the number of suppliers. Along with the PTT, energy, construction, trading companies, and banks have become active in telecommunications. The fast privatization of the British market has resulted in the British operators becoming active outside of the United Kingdom. Table 3.1 indicates that British Telecom (BT) and Vodafone are active in dozens of countries. The experience of these companies has often been the reason for offering mobile licenses to consortia with BT or Vodafone as stakeholder, by national governments in these countries. The consortia bidding for mobile licenses may vary from country to country. Vodafone and BT are each other's biggest competitors in the United Kingdom, but they participate together in consortia in France and until recently in Spain and Japan. The large telecommunications companies try to get a presence in as many countries as possible by acquisitions and licenses. This is referred to as the "global footprint." They also try to acquire stakes in regular telephony, Internet, and cable companies, found new companies, and set up alliances. The reasons for this are as follows:

- Purchasing advantages on acquiring mobile infrastructures and peripheral equipment. A select group of telecom suppliers dominate the market in innovation and price determination. Large operators have a strong relationship with suppliers such as Alcatel, Nokia, Siemens, and Ericsson.

- Customers from one country make use of an operator's preferred network when abroad. Operators will have their customers use their own network abroad. Also, placing a call from abroad to use a WAP service can be simplified and offered at cheaper rates if the operator can direct its customers to its own networks abroad.
- The development costs of the new services can be distributed among more countries. BT developed a mobile portal that can also be used by their subsidiaries in other countries; Vodafone did the same with Vizzavi.
- Volume gives operators a competitive advantage in negotiations on interconnection between networks and in dealing with calls on third-party networks.
- The cost associated with acquiring content can be earned back via more mobile customers. The operator obtains a preferential position in contract negotiations with strong content parties such as Disney, CNN, or Yahoo!.
- In a market in which maximum results have to be generated from new services, sharing experiences between operators can be beneficial.
- Many services will be offered via different devices and different networks. Owning a mobile network and a cable network gives the operator the option to offer customers a complete package of services.

It is clear that scalability is necessary for smaller telecom companies. They merge or try to create a "Pan-European footprint" with the support of American or Japanese telecom companies. The American SBC has a stake in the second French operator SFR, in Swiss number 2 Diax, and in Danish market leader Tele Danmark. KPN Mobile has joined forces with BellSouth and the Japanese NTT DoCoMo. Smaller mobile operators such as Tele Danmark, Swisscom, and Sonera try to realize expansion without taking ownership of a physical telephony network. Tele Danmark and Swisscom are owners of mobile service providers Talkline and Debitel. Sonera has founded Internet service providers in a number of countries and has introduced Internet portals together with Sonera Plaza and Sonera Zed. In Finland, Sonera will start a mobile trading place for consumers in cooperation with a large retailer. Its partner, the S-group, has more than 1,200 branches and can now offer services together with Sonera, varying from food to bikes to restaurant reservations to hockey tickets. If you can't beat them, join them!

Table 3.1 shows the footprint and other relevant activities of all the important mobile players. Most of the former monopolists have a strong position in telephony and Internet in their home market and have several vested interests in other countries. European regulations have been set to prevent operators from becoming monopolists. After the acquisition of the German mobile operator Mannesmann, Vodafone was not only the owner of the Vodafone network, the market leader in the United Kingdom, but also of the number-three UK operator, Orange, because Orange was owned by Mannesmann. Vodafone was forced to sell off one network. France Telecom's subsequent purchase of Orange UK lead to a similar situation in Belgium. France Telecom already owned Mobistar and also obtained a share in KPN Orange. KPN Mobile took over France Telecom's stake in KPN Orange Belgium. Due to the high expenses required for 3G, most of the operators are reconsidering their footprint strategy, either increasing their share to control the operator's operations or selling minority stakes in operators where a majority share cannot be obtained.

Table 3.1 European Telecom Players

Telecom Companies	Mobile Footprint	Other Relevant Information
Vodafone Mannesmann Airtouch (English-German-American)	UK (Vodafone), Germany (D2 Mannesmann), France (SFR), Italy (Omnitel), Spain (Airtel), Netherlands (Libertel), U.S. (Airtouch), Poland (Polkomtel), Portugal (Telecel), Romania (Mobifon), Sweden (Europolitan), Belgium (Proximus), Switzerland (Swisscom), Greece (Panafon), Ireland (Eircell), Malta, South Africa, Egypt, New Zealand, India, South Korea, Japan, Australia, China	Partnership with Vivendi for mobile portals

Table 3.1 European Telecom Players (Continued)

Telecom Companies	Mobile Footprint	Other Relevant Information
Orange (France Telecom) (French)	France (Orange), UK (Orange), Italy (Wind), Belgium (Mobistar), Denmark (Mobilix), Netherlands (Dutchtone), Portugal (Optimus), Poland (PTK), Slovenia (Globecast), Moldavia (Voxtel), Romania (Mobilrom), Greece (Panafon), Switzerland (Orange Comms), U.S., Latin America, South America (Argentina)	In the Netherlands, the owner of cable company Casema, and Internet suppliers Wannadoo and Euronet. Wannadoo, market leader in France, is also active in Belgium. Co-owner of NTL, a cable company in the UK.
British Telecom (English)	UK (Cellnet), Germany (Viag Interkom), France (SFR), Italy (Blu), Netherlands (Telfort), Portugal (Portugal Telecom), Ireland (Eset), South Korea, U.S., Canada, Singapore (Starhub), Hong Kong (Smartone)	BT Internet and Genie mobile (SMS) portal with 350,000 users also rolled out in Spain, Japan, and the Netherlands.
Deutsche Telecom (German)	Germany (D1), UK (One-to-One), Italy (Wind), U.S. (Voicestream), Austria (MaxMobil), Poland (Era GSM), Hungary (Matav/Westel), Czech Republic (Radiomobil), Croatia (Croatia Telecom), Canada (Microcell)	T-online is the largest European Internet provider, especially in Germany. It also owns Club Internet, the number-two market leader in France.
Telecom Italy (Italian)	Italy (TIM), France (Bouygues), Spain (Amena), Austria (Telekom Austria), Greece (Stet Hellas), Serbia (Telekom Serbia), Peru, Brazil, Argentina, Bolivia, Chile, Cuba	
Telefonica (Spanish)	Spain (T.Moviles), Portugal (Portugal Telekom), Morocco, Peru, Brazil, Chile, Cuba, Mexico, Argentina	Terra Lycos is the leading Internet provider in Spain and Latin America

Table 3.1 European Telecom Players (Continued)

Telecom Companies	Mobile Footprint	Other Relevant Information
KPN Mobile (Netherlands)	Netherlands, Germany (E-plus), Belgium (KPN Mobile), Hungary (Pannon), Ukraine (UMC), Indonesia (Telkomsel)	The largest Internet provider in the Netherlands together with Planet Internet, Het Net, and XS4all.
AT&T Wireless	U.S., Peru, Brazil, Argentina, Chile, Columbia	Also Internet and long-distance services.
SBC	Germany (Talkline), France (Cecetel), Netherlands (Ben), Belgium (Belgacom), U.S. (Cingular), Canada	

Service Providers

A service provider buys mobile minutes, subscriptions, and prepaid cards in bulk from mobile operators and sells them to mobile phone users. Not every country has service providers for mobile communications, due to differences in telecommunication laws. In some countries, the mobile operators entered into contracts with service providers to speed up the market development in that country. At the beginning of the mobile era, the price for mobile calls was high and the margin on minutes of call time was an important source of revenue. Service providers would get a percentage of those revenues. In the meantime, the prices of mobile call time minutes have decreased and therefore so have revenues.

Service providers purchase mobile call time minutes and do not receive compensation for receiving calls on mobile phones. Receiving calls is an important source of revenue for the operators. The arrival of prepaid schemes put further pressure on the margins for service providers. In most countries, service providers have therefore differentiated their product offering (e.g., by also offering Internet services) or have been taken over by other service providers or operators.

Another trend in the world of service providers is the arrival of virtual network operators. Virtual network operators go beyond the value chain of regular service providers. A virtual network operator issues SIM cards, without owning a network, and has its own services such as email notification, voice mail, or a mobile Internet portal. Branding is a very important issue for service providers and virtual network operators. The best example of a virtual network operator with strong branding is Virgin Mobile.

VIRGIN MOBILE CONQUERS THE GLOBAL MOBILE MARKET

Virgin is taking over the English mobile telecommunications market, after conquering music, airlines, and cola. After its introduction in November 1999, Virgin signed on 300,000 customers within seven months, reaching the 1 million mark during 2001. Virgin is the most well-known English brand and represents value in combination with good service. Virgin Mobile is positioned as simple and cheap, but forward-looking. Virgin Mobile uses the One-2-One network and outsources billing completely, focusing completely on marketing and distribution via the Virgin Megastores, Our Price shops, direct mail, and the Internet. As the Virgin Mobile brand is already known, a communications budget of $7 million was sufficient for the introduction period. Virgin announced plans to start similar services in other countries and has already started in Australia. The company has also started Virgin Mobile Asia with Singapore operator SingTel. The promise is made good by offering the customer a cheap and simple package. On purchase of a mobile phone, the customer is charged the price of the phone. As opposed to other suppliers, where the phone is almost free, Virgin mobile phones cost about $100. However, the rates are almost 30 percent lower and there is no monthly fee. Almost all Britons would pay less by using Virgin. Britons who already have a mobile phone can buy a Virgin SIM card for about

Figure 3.5
Virgin mobile Web site.

> $15 including 30 call-time minutes. This makes the switch to Virgin more interesting. The SIM card offers possibilities for ambitious services (see Figure 3.5). Every Virgin customer can use music notes as dial tones. The larger memory of a SIM card also allows the user to receive a three-page email via the mobile phone. Virgin has also introduced WAP on the SIM. As discussed in Chapter 1, a user can use WAP on a limited scale without having a WAP-enabled phone. Virgin also sells WAP phones in its stores and has adapted its outlets to multimedia stores. Virgin aims to create a global business based on the MVNO model. In Asia, Virgin aims for 10 operations in three to five years, servicing 6 to 10 million customers. Virgin also aims to launch services in the U.S. market soon (see *www.virginmobileusa.com*).

Network Suppliers

The network equipment supplier supplies those appliances needed to build and maintain a mobile network to the mobile operator. The five largest suppliers worldwide are Ericsson, Motorola, Lucent, Nortel, and Nokia. A mobile network is comprised of a radio network and a switching network. The radio network consists of base stations and antennas that connect the mobile device to the switching network. The switching network makes sure that the conversation is routed to the right destination. The switching network also keeps track of the location of a mobile device and whether or not it is switched on. In addition to this, the different suppliers offer so-called value-added services platforms, which enable mobile operators to offer value-added services such as voice mail or fax mail, SMS, and voice dialing. Well-known suppliers of such services are CMG, Glenayre, Unisys, Comverse, Bright, and Wildfire.

The mobile operator depends on the supplier of the network equipment for the quality and the speed at which it can supply the new services to the market. A mobile operator cannot switch network suppliers easily due to the enormous investment in its network. The competition in this field takes place out of sight of the mobile user, but the amounts of money involved are enormous.

Mobile Telephone Suppliers

The mobile telephone is the fashion item of choice. In the beginning, phones were big and bulky and only available in black and gray. The current models complement the lifestyle of the customer. With removable faceplates it is possible to color-coordinate your phone with your outfit. The most important suppliers worldwide are Nokia, Motorola, Ericsson, Siemens, and Panasonic. Worldwide Nokia, Ericsson, and Motorola together make more than half of all mobile phones sold. In 2000, more than 412 million mobile phones were sold.

Different Roles in Mobile Communication

Figure 3.6
The Samsung SPH-WP10: Watch and mobile phone in one.

The competition between device suppliers is intense. The important selection criteria for the customers besides price (often subsidized by the mobile operator) are size, weight, battery, functions, and appearance. Size and weight are important for user comfort. In this area there has been tremendous progress in the last few years. The first mobile telephones were built into the car and were the size of a car battery. Today the size and weight of a phone is minimized. Samsung introduced a watch that could also be used as a phone (see Figure 3.6).

The battery can make or break the success of a telephone. Many users want to call for long periods of time and always want the phones to be switched on. An uncharged battery is annoying. Many manufacturers use lithium-ion batteries, which allow for much longer call times and standby possibilities without increasing the weight of the phone.

Functions and accessories are more important for the experienced caller. Is there a vibrating mode, can I select the dial tone, can I send SMS and pictures, and can I use WAP? This is just a small selection of questions asked of mobile phone salespeople. In the shape of accessories, faceplates in multiple colors are an important weapon in competition. Also, the availability and quality of headsets and car kits is frequently a part of the selection process. For those who already have a mobile phone, the possibility of reusing accessories—especially car kits—is important, and a smart way of creating customer loyalty!

Design and user friendliness are the final important criteria. Does the user like the design, is he or she pleased with the user interface, and does it fit with his or her lifestyle? Films and television can play an important role. Ericsson used a James Bond movie for an advertisement, in which the hero used a super-small Ericsson telephone.

MOBILE PHONE USAGE

In Europe, traveling salespeople were the first users of mobile telephony. After the price for mobile telephony decreased as a result of new competition, more and more consumers started to use it and the penetration doubled year after year in some countries, stabilizing at around 70 percent. The mobile telephone is mostly used for calling, receiving calls, and voice mail.

The use of mobile phones differs between user groups. In general, businesspeople call more often than other consumers do and they often call during the day. They also use the mobile telephone much more during workdays than on the weekends and more often abroad. Business callers are less price-sensitive than other consumers are because they do not have to pay their own phone bills and because the use of mobile phones makes their work more efficient and easy. A mobile phone goes hand-in-hand with certain jobs. This applies to salespeople, management, and off-site employees. These days, we see that employees who are in the office five days a week often have a company mobile phone. Mobile phones have now become a fringe benefit that can be used for personal and business matters.

General consumers often call after work and on the weekends and in general much less than business users. The prepaid system is very popular with consumers because the costs are easier to control and there is no monthly fee. The customer will not have to undergo a credit check for prepaid as opposed to subscriptions. This way, young people who do not have cosigners still have the ability to get a mobile phone. Scratch cards with mobile calling minutes have become popular gifts for birthdays and other occasions. Next to the use of mobile phones for calling and receiving calls, SMS has become a popular service among younger users.

Finland is far ahead of other countries when it comes to SMS, with usage at more than 65 percent SMS and 35 percent speech. SMS messages are very popular in the youth segment. The 75 percent penetration of Nokia telephones in Finland is part of the explanation, as these make sending SMS messages relatively simple, especially retrieving and receiving new dialtones via SMS. Mato Valtonen, founder of Finnish WAPit, indicates that the killer applications in Finland are not stock quotes or headlines, but jokes. Of the 200 different SMS services that WAPit offers, dumb-blonde and lawyer jokes, horoscopes, and biorythms are most popular. Anonymous chatting using SMS via mobile phones is very popular also.

In December 2000, 15 billion SMS messages were sent worldwide. The average SMS traffic per GSM customer has grown from 0.4 in 1995 to an average of 35 messages per GSM customer per month by the end of December 2000 (see Figure 3.7).

The increase of SMS use is an indication that users no longer see their mobile phone as an instrument for speech only, but also as an instrument to send messages and obtain information. In the United Kingdom, the national blood bank donor service is currently using SMS to generate calls to young donors and regularly remind volunteers of appointments. In some European countries, Muslims use SMS services to notify daily calls to prayer.

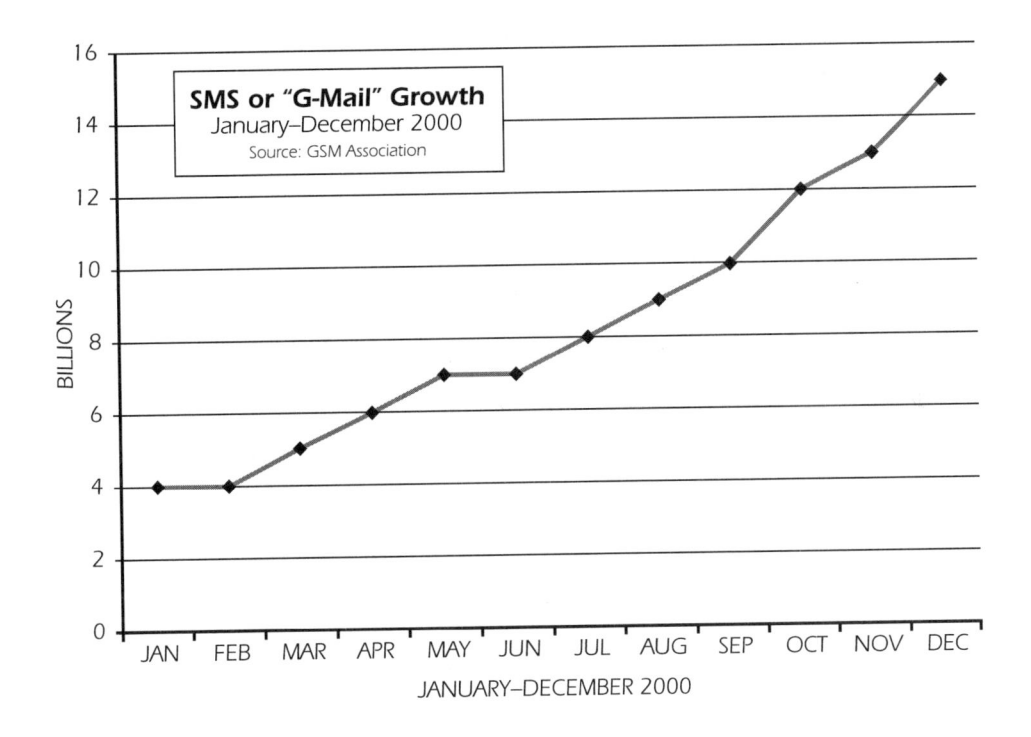

SMS or "G-Mail" Growth
January–December 2000
Source: GSM Association

Figure 3.7
Growth in SMS traffic worldwide in the year 2000.

MOBILE TEXT CHAT

Mobile text chat (see Figure 3.8) is a service with which mobile users can send SMS messages to an abbreviated number. These messages are then displayed on teletext. As more people are sending messages, a chat occurs, in which participants to the chat react to the messages of other users. Mobillion introduced this service in the Netherlands in cooperation with SBS Broadcasting. This service was introduced in June 1999 and has drawn a wide user base. It even resulted in a marriage!

Figure 3.8
Chatting using the mobile phone and teletext.

MOBILE COMPANY NETWORKS

Business users of mobile telephones are often not the people paying for these services. As a result, different concerns arise. The user wants to use the mobile phone as often as possible (even if he or she is in the office), as the mobile phone is always at hand. The user will also use the phone at night and on the weekends for personal use, even if it is just so he or she can be reached. The company telecom manager wants to control the use of mobile phones, in respect to cost and technical support toward employees. The responsibility for cost control can also lie with the employee's manager.

Mobile operators are alert to these problems and offer special services to companies. These services are called mobile VPN (virtual private network) service. Employees can call each other at lower rates using internal extensions. This goes for mobile users calling colleagues as well as employees in the office calling colleagues on the road. Some mobile VPN services allow the telecom manager to install a connection profile for every user. With every call to the user, the connection profile of that user is used; for example, first the land line is dialed, then the mobile line, and finally the secretary's number. The telecom manager can program restrictions and create so-called white and black lists. In this way the mobile phone can be disconnected after work hours. It can also be decided that certain employees can only dial internal numbers from their mobile phones. The cost per user can then be allocated properly. This also limits excessive calling.

In the business market, prepaid phones are often given to those employees who seldom need to use a mobile phone. The employer can give all employees a certain amount in prepaid minutes for a certain period of time, after which they will have to pay for their calls themselves.

Mobile operators also offer special services to business customers supporting use of mobile data. This enables users to access the corporate network with a laptop using GSM's limited 9.6 kbps data speed. Also, SMS offers for companies sending large amounts of SMS messages to employers or customers are specially packaged.

4 The Internet

In this chapter...

- Different Roles in Service Provision   103
- Internet Usage   110
- Company Networks   112

Similar to mobile telephony, the Internet has not been around for a very long time, and it has only been very recently that the Internet has taken off for the majority of companies as well as the mass market of home users. The Internet started out as a network for the U.S. defense industry and later as a network between universities. The Internet was mainly used for three applications: email, newsgroups, and FTP (file transfer protocol, sending files via the network). With the invention of the World Wide Web by Tim Berners-Lee, it became possible to create a home page with text and graphics. Users with a computer and a browser could view this home page and download files from it easily. The World Wide Web and decreasing prices for computers and telecommunication connections allowed for a breakthrough of the Internet among consumers and companies. The United States takes the lead in this area. In the United States in March 2000, almost 50 percent of the population had access to the Internet and almost 15 percent had made at least one purchase over the Internet. By comparison, in Europe, 34 percent of the population had access to the Internet and 8 percent had made a purchase. The average American surfs the Net about one hour each day compared to 15 minutes for the average European.

In the United States a new industry branch of Internet companies appeared. Many of these companies are located in the San Francisco area, in the so-called Silicon Valley. The Internet industry is known for its dynamics. New companies are started every day. Even if a large number of them disappear from the market, the success stories appeal to the imagination. People like Jim Clark (Silicon Graphics and Netscape) and Jeff Bezos (Amazon.com) have succeeded in turning their attic companies into billion-dollar corporations. The dynamics of this industry are noticeable in the rapid time to market of new products and services and the building of partnerships, mergers, and acquisitions of companies. The company culture in most Internet companies differs from the more traditional industries. This is due to the fact that many companies have started from a hobby. Think of Bill Gates (Microsoft) and Steve Jobs (Apple). The casual dress code and the lack of status symbols are a result. The time between being in and out of fashion can be short, so people are always alert to developments and they keep busy improving the current product and the development of new products. Cooperation with other companies is different here than in other industries. It is common to exchange ideas and to enter into strategic alliances.

The real believers in the Internet even mention a "new economy" that exists due to the Internet, in which the current laws of supply and demand would no longer be valid. Many Internet companies manage to be popular at the stock exchange although they do not make a profit. On the one hand, this is logical, as a company needs to keep up investments to create a good market position in a growing market. On the other hand, the stock value is then purely based on expectations. Behind many companies there is no good business model and it remains to be seen if these companies will ever be profitable. After the enormous success of Internet stocks in 1999, they generally took a sharp fall in 2000 because stockholders lost trust in the eventual profitability of those companies.

DIFFERENT ROLES IN SERVICE PROVISION

In the Internet industry, there are many distinguishing roles: the Internet access and service provider, software suppliers, hardware suppliers, network suppliers, portal providers, content providers, merchants, and advertisers—terminologies in need of further explanation. It is too much to explain the Internet in great detail here. The interested reader is referred to books focused on the Internet.

Internet Access and Service Providers

The Internet access provider delivers access to the Internet. The user connects to the Internet access provider by dialing a local number with his or her modem. Internet access providers offer local access numbers so the user can connect with lower telephone rates. The Internet access provider manages a modem pool and a router, which is connected to the Internet with a large-capacity connection. Besides local access via analog modems over the regular phone line, a user can access his or her Internet access provider via ISDN, xDSL, or cable (depending, of course, on the options offered). Many Internet access providers call themselves Internet service providers, as they also offer other services besides Internet access, such as email, Web proxy, design, hosting of Web sites, and a help desk. For companies, they also offer services such as rental access, domain name registration, and hosting of the company Web server.

The most well-known Internet service providers are America Online and UUnet. Most Internet service providers started with a business model based on subscription revenue. The user pays a fixed subscription fee or a price per hour for connection to the Internet. Later, new business models were introduced. In some countries, the telephone operator offers a kickback fee to Internet service providers, a small percentage of the revenue the telephone operator makes, for the time users were connected to the Internet service provider. Another source of income for Internet service providers are advertisement banners, provided of course that they offered their users an interesting home page to start surfing from (see Figure 4.1). The kickback fee and advertisement revenues made some Internet service providers offer their services for free. In most cases, however, this is a very difficult business model to make profits from.

Some Internet service providers offer flat-fee subscriptions that include local telephone or cable costs. The user pays a fixed amount per month for unlimited Internet access. Quite a few cable companies use this method. The big disadvantage of this model is that if you offer unlimited access, you will attract users that will make use of

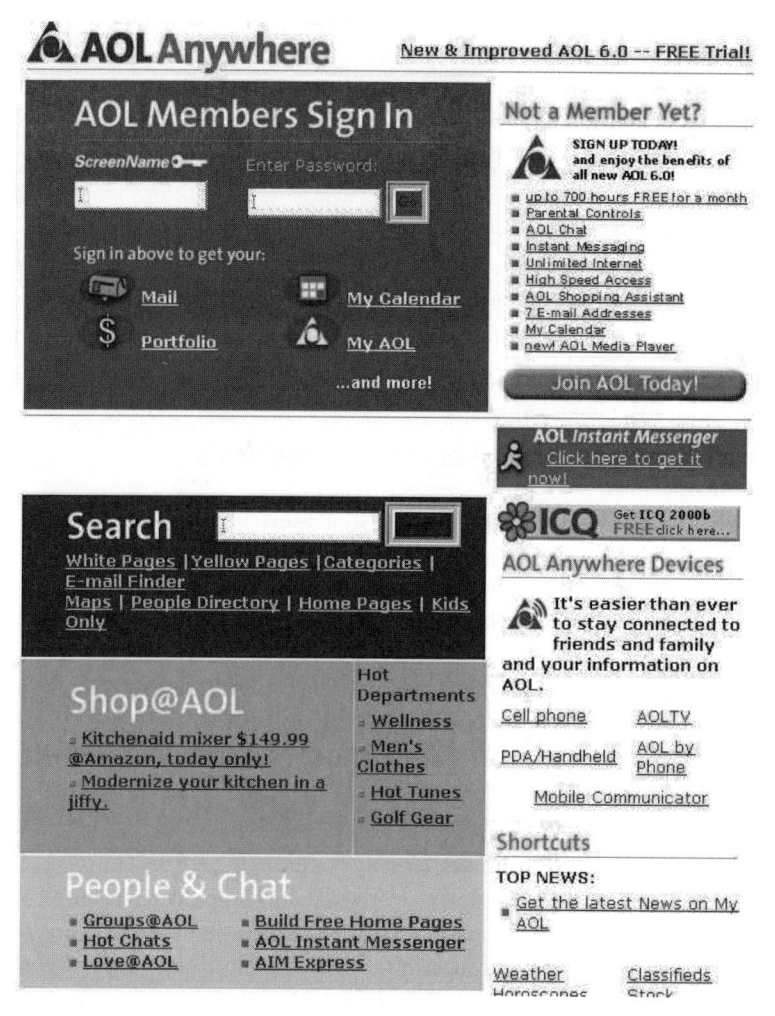

Figure 4.1
Home page of AOL.

it. Often, the predicted use per user per month was surpassed a few times. Loading MP3 music clips at night is a well-used application among these users. Another big disadvantage of this model is that by attracting frequent users, the overall quality of the service decreased because of the heavy load.

In the selection process for an Internet service provider, price, quality, and service play an important role. Free Internet service providers often appear to be free, but they are not. The user still pays phone charges for dialing in. The user often has to give much more personal information than with other Internet service providers and the Internet service providers often sell this information for marketing purposes. Other services offered are not free, like help-desk use. Many free Internet providers do not offer the network quality and speed that paid Internet service providers offer as a result of the great number of subscribers. For the speed of an Internet connection, not only the bandwidth to the Internet service provider is important, but also the router capacity of the Internet service provider and its bandwidth toward the Internet. In many computer magazines, tests are published to draw attention to the quality and differences in service levels between the different providers.

The revenues from the kickback reimbursement are not fixed. Telecommunication suppliers do not have to share their revenue with ISPs. The battle for the customer intensifies. Besides free Internet access, most Internet users will not have to pay local phone rates in the future. This is currently the case in most European countries. The supply of fast Internet access has to compensate for the loss in revenue, but sizable investments are needed. The other sources are Web advertising and electronic commerce. In 1999, this did not amount to much. For example, 3 percent of service provider World Online's revenue came from Web advertising and 9 percent came from electronic commerce. The following years showed no increase in revenues from Web advertising due to the changing Internet climate.

Software Suppliers

The browser supplier offers user software needed to view Web sites created with HTML. The most important browsers are Netscape Navigator and Microsoft Internet Explorer. Netscape is mainly used in non-Microsoft environments, such as Linux and Apple Macintosh, but can also be used for Microsoft Windows. Microsoft supplies Microsoft Internet Explorer free with Windows software, which was one of the issues in the Microsoft antitrust lawsuit. The development of browsers therefore never became a profitable business and browser suppliers often make money in different ways: Netscape from the sales of Web server software and Microsoft from the sales of Windows software. Besides the browser suppliers, there are many other software suppliers that facilitate the value chain with their products. Some software suppliers aim purely at the support of portal suppliers, content providers, merchants, or advertisers with the construction of Web site software, software to register user behavior, payment software, or firewall software. Other software suppliers target the Internet service provider or the network supplier. ERP and EDI suppliers adapt their traditional systems. SAP has launched MySAP.com and Peoplesoft has introduced E-procurement as an Internet purchasing solution. The business-to-business markets appear to be dominated by Oracle, which can finally surpass IBM and Microsoft. There are dozens of smaller companies, such as Commerce One and Ariba, that want their share of the market, too.

Hardware Suppliers

In the world of the Internet, there are three important types of hardware suppliers: those that supply hardware to the user, the information provider, and the Internet service provider. In 2000, 134.7 million PCs were sold worldwide, according to Gartner. The most important PC vendors worldwide were Compaq, Dell, and Hewlett-Packard. The Internet is a packet-switched network of routers. Cisco is the market leader in Internet routers, the basis of the Internet.

Network Suppliers

Internet routers are connected with data connections. Telephone operators often supply these connections. Well-known operators are AT&T, Sprint, BT, MCI Worldcom, Colt, Telefonica, and Global Crossing. These telecom operators earn their money with transportation of the enormous quantity of data and speech traffic at relatively low margins. Most of these companies built fiberglass rings in Europe and the United States, in and around cities, which companies and Internet service providers can connect to. There has also been an initiative to build a fiberglass ring around Africa with branches to all African capitals. With this, Africa can also be connected to these world-encompassing networks.

Portals

A portal, as the name indicates, is a port of entry from which the user starts the search for information on the Internet. A portal can be aimed at an area of interest, for example, books, news, or stock information. A portal can also be a page with a search engine that will help the user find the information needed. Internet service providers often supply their customers with a starting page or portal from which they can start their search. The most important source of revenue for portals is advertising. Advertising revenue is higher if more is known about the visitors and the advertisements can be targeted at a better segmented group. For advertisers it is more beneficial if their ad is exposed to the audience that is searching for this type or similar type of service.

The most well-known international portals are Yahoo!, AltaVista (see Figure 4.2), MSN, Macromedia, ICQ (online communication with friends), MP3, and tucows. It is often difficult to distinguish between a portal and a regular Web site. Both can have their own content and link to other sites. The main difference probably lies in the business model. Is the site intended to sell something (e.g., newspapers, books, magazines, airline tickets, products, or services) or is it intended to create customer loyalty for the portal, and to maybe gain information on these customers with the intention of making an interesting offer to advertisers? The latter we call a portal. The word "maybe" in the preceding sentence indicates that not all portals have a business

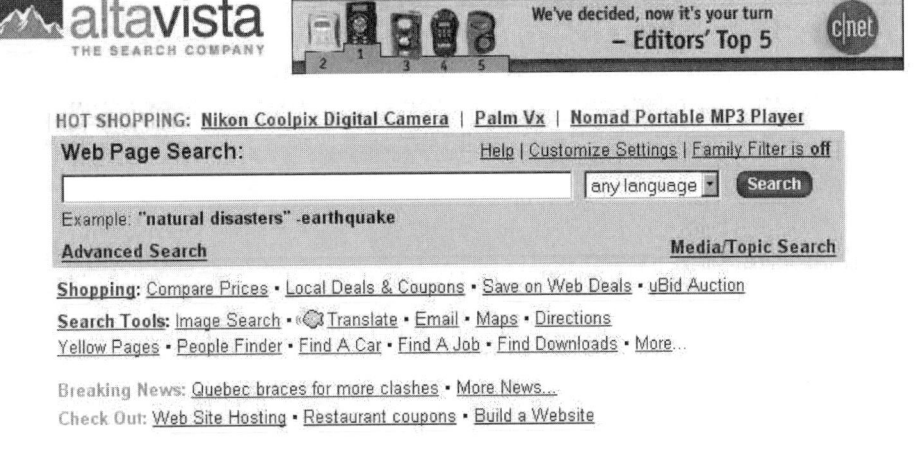

Figure 4.2
Altavista search portal.

objective. There are many portals based on interests and information sites that do not have a business objective as a base. There are other portals or comparison sites that help visitors with the selection of a product. The business case of these portals is based on revenue from Web avertising and transaction reimbursements.

Content Providers, Merchants, and Advertisers

A content provider has a Web site on which it offers information, either free or at a price. In general, it is difficult for a content provider to get money for Internet information, unless the information is highly specialized and uses the Internet as a cheaper means of transport. Why then are there so many content providers? The first reason is that many content providers on the Internet give a sample or a complete overview of the actual content (newspaper, magazine, TV, or research report). Examples of this are newspapers like the *International Herald Tribune* and the *New York Times* (see Figure 4.3), magazines like *Vogue*, television stations like CNN, and research company sites such as Gartner and Forrester. The second reason is that with many sites, not the buyer but the supplier pays to be hosted at the site. A third reason is that many companies and individuals want to host their own information on the Internet. This can be in the shape of building Web sites and contributing information to other sites. A famous example of the latter is Amazon.com, where buyers can write their own book reviews.

Figure 4.3
The *New York Times* on the Web.

  Merchants have Web sites where they offer their products and services. The difference between content providers and merchants is sometimes vague, as some suppliers make all information about their products available but the purchase is not possible on the Internet. Well-known merchants for consumers are Amazon.com (books and CDs) and eBay (auction of goods).

  Existing "brick-and-mortar" retailers do not give way to their virtual competitors easily. They start their own sites, some very successfully. In the United States, Toys 'R Us opened its own site as a response to eToys.com, which competed with Toys 'R Us stores. Customers search on the Internet for parties they can trust. Dutch catalog merchant Wehkamp realized $10 million in revenue via the Internet in 1999, 41 percent coming from clothing. Experience as a catalog sales company has turned out to be a great benefit for Wehkamp in setting up an online store. Levi's has omitted e-commerce from its new strategy. Channel conflicts and high operating costs of its site have forced Levi's to stop its e-commerce activities. The bankruptcy of Boo.com indicates that the trees do not grow to the heavens for Internet merchants. Large expenses of more than a million dollars per day for new technology and marketing did not result in sufficient revenue. This has shaken entrepreneurs and investors, supported by research companies and banks that say that a large number of these companies will disappear.

Most money on the Internet is made in the business-to-business market. A well-known example of this is Dell, which now offers its computers via the Internet with special software that allows the customer to determine the configuration (see Figure 4.4). This allows Dell to supply directly to dealers and companies and to prevent administrative hassles. Fifty percent of Dell's total sales in 2000 came from the Internet ($40 million per day in 2000, as opposed to $14 million per day in 1999).

Purchases made via the Internet will increase as well. The Internet plays the part EDI has fulfilled for years. The energy auction Buying Power offers companies the opportunity to bundle purchasing power to obtain a better price. A number of airlines have set up a trading marketplace, initiated by Boeing, for the purchase of kerosene and parts. General Motors has also set up a purchasing portal in close cooperation with Ford and Daimler Chrysler, to which 40,000 suppliers are linked. Sears and Car-

Figure 4.4
Dell's Web site.

refour have transferred their entire purchasing process to a purchasing site. In the meantime, other retail chains, such as Metro A.G. and J. Sainsburry, have joined. The purchasing sites increase transparency, resulting in savings with regard to purchasing costs. This also reduces the cost of ordering in the purchasing process.

Advertisers place their ads on a third-party Web site, in the form of a banner or a button. The reason for this could be to draw the Web surfer to the site or place an order right away. Advertisers pay either by page view (the number of times the ad is shown on the site) or per click through (the number of times the ad is clicked to visit the advertiser's site). There are other options, by which the merchant or advertiser pays a percentage of the sales price if the user purchases the product via the Internet. Only a small part of the average company's communication budget (about 5 percent) is spent on online advertising. In Europe, this is still $200 million per year. In the next few years, the market for online advertising will double in Europe, according to AOL.

Transaction providers take over the transaction from the merchant. The transaction provider provides a secure transaction of the payment details and also finalizes payment to the merchant. Many merchants prefer to keep these processes in house, but some outsource to specialized parties. In the Netherlands, Bibit provides a payment service to merchants. The customer selects the product he or she wants to buy on the merchant's Web site. Next, the customer is connected with Bibit's payment service, where he or she can select a method of payment, such as credit card, check, or debit card. On payment, the merchant is notified, and the product can be delivered.

Logistic service suppliers make the link between their Web site and logistics: e-fulfillment. All packages ordered via the Internet need to be delivered. And not just in two weeks, but sometimes tomorrow by 6:00 p.m.! The European Express Organization expects an annual growth of 18 percent in the European express industry. TPG specially developed a system, Internetdienstverlening@TNT, to link order entry, inventory, and delivery. UPS, FedEx, and DHL have similar initiatives.

INTERNET USAGE

According to Nielsen, in the second quarter of 2000, 269 million people across 20 countries in Europe, Asia Pacific, and North America had Internet access from a home PC. According to eTForecasts, the total number of Internet users was 375 million at the end of 2000 (see Table 4.1).

In Europe, average Internet penetration was 40 percent (representing 127 million persons) among inhabitants over the age of 15 in the third quarter of 2000 (Pro Active International). The penetration is highest in the northern European countries of Norway (69%), Sweden (67%), and the Netherlands (60%), and lowest in the southern European countries of Italy (35%), Portugal (17%), and Spain (13%).

Table 4.1 Top 15 Nations in Internet Use at Year End 2000

Rank	Nation	Internet Users (millions)	Share %
1	United States	135.7	36.20
2	Japan	26.9	7.18
3	Germany	19.1	5.10
4	UK	17.9	4.77
5	China	15.8	4.20
6	Canada	15.2	4.05
7	South Korea	14.8	3.95
8	Italy	11.6	3.08
9	Brazil	10.6	2.84
10	France	9.0	2.39
11	Australia	8.1	2.16
12	Russia	6.6	1.77
13	Taiwan	6.5	1.73
14	Netherlands	5.4	1.45
15	Spain	5.2	1.39
	Worldwide total	374.9	100.00

Source: eTForecasts

The average use per active user is very different from country to country. According to Nielsen, in January 2001, South Koreans were the most active users, with an average of 16 hours per month spent on the Internet, compared to 10 hours in the United States and 6 hours in the United Kingdom. The South Koreans looked on average at 27 different sites, compared to 10 sites for users in the United States, and 18 for users in the United Kingdom. It is probable that since the United States is a relatively mature market for the Internet, U.S. users visit less sites, because they know where to go for their content, whereas Europeans and Asians are still exploring the Internet for content.

Nielsen compared Internet use in Europe with the United States and found that in both regions search engine portals and telecom Internet services are the two categories that attract the most traffic. In the United States, entertainment and shopping are the most popular categories after those, whereas in Europe, personal business electronics and software news and information rank third and fourth.

Email is the most-used Internet-related activity. Chatting, newsgroups, and downloading MP3 (music) files are also very popular applications. Most of the time spent on the Internet is at the expense of time spent watching television.

Purchasing online is a growing market. The top items purchased online are books, CDs, computer software, and travel-related services. Forrester expects a $7 trillion market worldwide by 2004, with North America ($3.5 trillion), Asia Pacific ($1.6 trillion), and Europe ($1.5 trillion) being the most important regions.

Aside from buying things on the Internet, many customers use the Internet to research products they are going to buy in the real world, for example, in the case of automobiles, computer hardware, and travel. Reasons for buying offline while researching online are to see the product before you buy and to have it immediately instead of having it delivered.

Online sales is a fast-growing market, especially in business-to-business sales. In the Netherlands, business-to-business sales were twice as high as business-to-consumer sales in 2000.

COMPANY NETWORKS

Even more so than with regular and mobile telephony, companies using Internet technology have a need for a secure environment shielded from the outside Internet. This is called an intranet. Intranets offer powerful opportunities to improve internal communication and to give employees access to the most up-to-date information, such as product information, prices, stock levels, and company news. The importance of an intercompany phone directory should also not be underestimated. An intranet demands a different attitude toward work and employees. Information does not automatically take the format of internal telephone directories and price lists on the desk. The employee has to actively search for the information he or she needs for the job.

Intranets are often created as a hobby by a smart IT employee in the company. To use the opportunities supplied by the intranet, clear business objectives and sound information management are necessary. Employees have to be educated in the new processes.

Extranets are secure networks based on Internet technology, but in this case the user group consists of employees, customers, or suppliers. Various larger companies have created extranets with their suppliers, where they place the orders with quality standards and price indications so that suppliers can make offers.

5 Mobile Internet

In this chapter...

- Mobile Internet outside Europe   117
- Different Roles with Mobile Internet   122
- The Fight for Market Share for Hybrid Devices   124
- Conflicts in the Value Chain   143

With Mobile Internet we refer to all the forms of mobile access to the Internet. Besides WAP and WAP-like services such as i-mode, we also considered alternatives to WAP, such as mobile access to the Internet via the laptop with a "normal" HTML browser. The combination of WAP browser or HTML browser with the new generation of mobile networks will replace the objections raised against WAP. Presented by the industry as the mobile Internet, WAP was initially disappointing to users. After the initial hype about WAP at launch, the press became very negative. However, according to Strand Consult, 61 percent of WAP users are satisfied. Telecomy claims 71 percent of WAP users say WAP is meeting or exceeding their expectations. Technologies like GPRS and UMTS create bandwidth and therefore speed. Just as with the PC, applications will improve step by step, using the increasing bandwidth. According to Pro Active Research, a year after launch, 4 percent of Dutch Internet users used WAP, although 90 percent had a mobile phone. This seems to be a low figure, but compared to the take-off of SMS, penetration is high. SMS penetration was only 6 percent in the Netherlands three years after launch. Prospects are quite good: more than 14 percent of mobile phone customers started to use WAP within 12 months, and 28 percent would eventually start using WAP. More than 50 percent of Internet users said that they do not use WAP because their mobile phones do not support it. The following factors will play an important part in the distribution and use of the mobile Internet:

- *Terminals*. The user needs a new terminal for mobile Internet. At this stage, dozens of terminals suitable for WAP are available, although it took longer than most industry experts expected. According to Nokia, 50 million WAP handsets were in use at the end of 2000. WAP became more and more a standard feature of new handset models. As active users change terminals on average once every 18 months, it will take a while before all targeted mobile Internet users have a suitable terminal at their disposal. Research company Forrester expects that in 2002, 107 million Europeans will own a WAP phone, of which half (14 percent of all Europeans) will actually use mobile Internet (see Figure 5.1). By then, Nokia and Ericsson expect that no devices without browsers will be produced. The availability of terminals has been a critical factor for every new technology launch. The rollouts of 3G in Japan and GPRS in Europe have suffered from low handset availability as well.

Internet Users/Wireless Users
(millions)

	2000	2002	2005
United States			
Internet Users	135	169	214
Wireless Internet Users	2	18	83
Worldwide			
Internet Users	414	673	1,174
Wireless Internet Users	40	225	730
Western Europe			
Internet Users	95	148	246
Wireless Internet Users	7	59	168

Source: eTForecasts

Figure 5.1
Expected number of Internet and wireless Internet users.

- *Supply of services.* Initially the possibility of using mobile Internet services will be an important reason to buy a WAP-enabled phone (see Figure 5.2). Once WAP has become a standard feature on mobile terminals (see Figure 5.3), the services provided will determine whether the users employ mobile Internet. The services offered via mobile Internet have to offer the user benefits. It has to be easier to use than other services, more fun to use, or less expensive. Currently, tens of thousands of developers are creating applications and content. In March 2001, there were 10,000 WAP sites from 95 countries according to Cellmania, and 7.8 million WAP-readable pages according to Pinpoint Networks. These numbers grow every day. There are very successful WAP sites: The Digital Bridges WAP game site had 23 million hits in six months, and DHL Worldwide Express had 250,000 hits on their WAP tracking site in the first year (compared to 25,000 hits on their Web site in the first year).

- *Availability of services for early adopters.* A Nokia study on value-added service users indicates three groups as the first adopters of mobile services:

    - Teenagers (younger than 18 years old)

- Students (between 19 and 25 years old)
- Young businesspeople (between 25 and 35 years old)

The introduction of @info was mainly aimed at this latter group, because the user required a subscription. Most teenagers and students use prepaid services, so they were not able to use WAP at first.

The Bureau of Social and Cultural Statistics in the Netherlands has conducted research in cooperation with the University of Utrecht concerning the digitalization of society. In this study, the differences in use and ownership of ICT products, such as television, video, and personal means of communication such as telephones, Internet, and email for various population groups were researched. This study showed that in general, men, youth, people with higher levels of education, and people with higher incomes take the lead when it comes to owning ICT products. Single women, people age 65 and over, people with lower levels of education, and people with lower incomes and the unemployed lag behind when it comes to IT. This will not be different with mobile Internet. The first users will mainly be young men with a higher income who also own and use other ICT products.

According to the Yankee Group, there were 220 million digital mobile telephone users and 150 million Internet users in 1999. In 2004, the Yankee Group expects more than half a billion Internet users, and roughly 1 billion digital mobile telephone users. IDC researchers made a similar estimation. For mobile Internet, 48 million users are expected at the end of 2002, and 204 million users for 2005. The expectation is that in

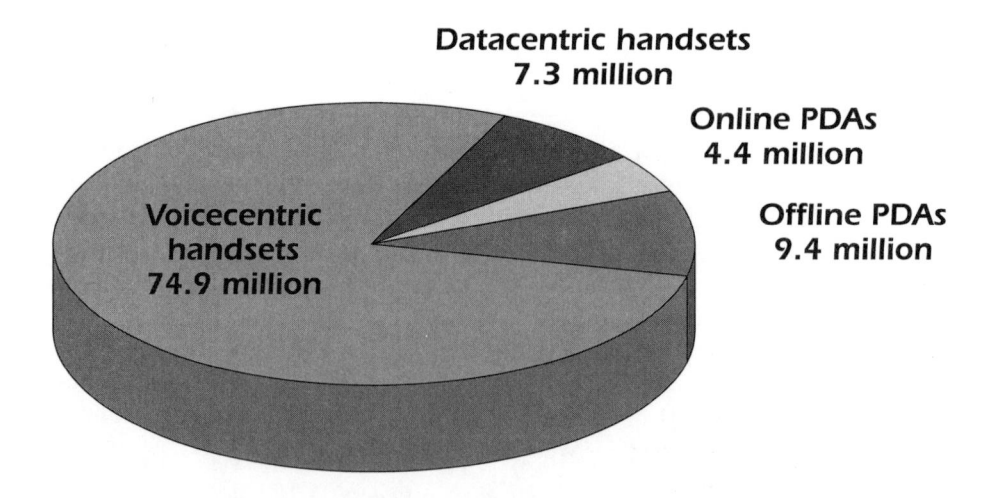

Datacentric handsets
7.3 million

Online PDAs
4.4 million

Voicecentric handsets
74.9 million

Offline PDAs
9.4 million

Figure 5.2
Expected use of different types of terminals by mobile Internet users in 2005.

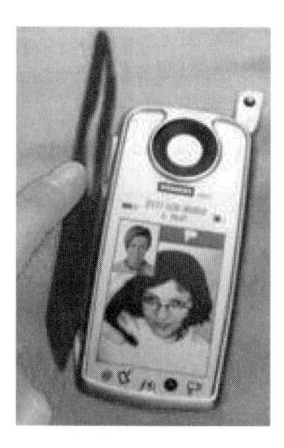

Figure 5.3
Siemens prototype, supporting video conferencing.

2004, one third of all Europeans will use mobile devices to access the Internet. This is an enormous market that offers many opportunities for both existing parties and new players.

In the business-to-business market, we can distinguish three types of organizations with different needs regarding mobile Internet:

- Sales-driven organizations like banks and production companies
- Service providers such as consulting firms and systems companies
- Logistic service providers such as taxi companies and couriers

Depending on the size of the organization and the need for services, specific mobile applications or integration in existing environments will be selected.

MOBILE INTERNET OUTSIDE EUROPE

Starting in 1999, mobile Internet was introduced all over the world. Although most of this book is dedicated to the Western world with a focus on Europe, Japan is currently the leading country in mobile Internet. Although their services and portal outlines have a lot in common, the Japanese have succeeded in marketing the services very well to a large group of users.

i-mode in Japan

Mobile telephony is very popular in Japan, especially with young people. As a result of the shortage of houses and high rents, many young people still live with their parents. Their mobile phone is their lifeline to friends. In Japan, the price of using a mobile phone is lower than that of using a land line. Japan has more mobile callers (60 million) than regular land-line connections. It is therefore not remarkable that only 20 million of the 125 million Japanese (13 percent) have a PC at home. Only 4 million Japanese access the Internet from home, and 16 million have Internet access at the office.

Various operators are active in the Japanese mobile market. Since 1999, IDO and DDI have offered WAP services with platforms supplied by American Openwave. The services, called EZ Access and EZ Web, had 3.5 million active users in February 2001. They take second place after NTT DoCoMo (the abbreviation stands for DO COmmunicate over the MObile network). DoCoMo is the largest mobile operator in Japan, with a market share of about 60 percent. DoCoMo launched mobile Internet in February 1999, in cooperation with major handset partners. DoCoMo had more than 19 million mobile Internet users as of 2001, welcoming more than 1 million new i-mode (the DoCoMo portal) customers each month. i-mode has become standard because 90 percent of new DoCoMo customers sign up for i-mode. Expectations are that soon 40 to 50 million DoCoMo customers will use i-mode. DoCoMo offers a wide variety of manageable mobile Internet telephones—small and lightweight, with features such as voice recognition, radio, and color screens. In contradiction to operators in the rest of the world, DoCoMo has a strong influence on the functionality and design of the phones, spending more than $800 million on research and employing a dedicated team of 900 engineers. It cooperates with Japanese handset manufacturers like NEC, Mitsubishi, Panasonic, and Fujitsu, but also with Nokia and Ericsson. Sony did not join i-mode at first, but entered the scene later. The next steps are music on mobile phones and a connection between i-mode and the Sony Playstation.

The success of i-mode was so overwhelming during 2000 that the email server regularly reached maximum capacity and couldn't be accessed. For that reason, the sales of new i-mode telephones were not really pushed for a few months. i-mode has not been positioned by DoCoMo as a high-tech gadget, but as an easy-to-use service. Half of the i-mode users are in their early 20s, and 30 percent of the users are female (Figure 5.4).

Next to the ample choice of mobile telephones, the success of i-mode can also be attributed to the large number of available Web sites, much more than are available from the competitor's portals. DoCoMo has chosen to access as many services as possible and cooperates with a large number of partners. In January 2001, the i-mode portal offered access to almost 800 partners and 38,000 sites that offer news services, weather forecasts, CDs, books, recipes, tickets, games, telephone directories, search engines, rental cars, real estate information, horoscopes, car navigation, and so on.

Mobile Internet outside Europe 119

Figure 5.4
Over one third of i-mode users are female.

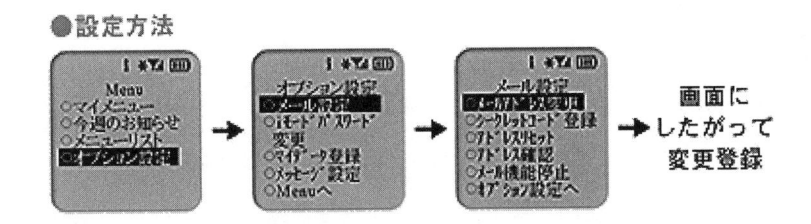

Figure 5.5
Example of an i-mode service.

One hundred sixty banks have joined together with credit card companies, insurance companies, airlines, and newspapers (Figure 5.5). In compliance with Japanese tradition, even karaoke bars are available. International companies like CitiBank, Disney, CNN, Bloomberg, Universal, Dow Jones, and Northwest Airlines joined i-mode. Downloading pictures, logos, and jingles is the most popular service. Another 15,000 sites are available for mobile Internet via i-mode. Bandai, the inventor of the Tamagotchi, sends a daily cartoon (Figure 5.6) to 1 million Japanese mobile phones for $1 a month. The Sakure Bank has more than 45,000 customers that purchase products via mobile phone or make payments via the i-mode portal. This service has more users than the home banking service offered by Sakure via the PC and Internet. Forty percent of the stockbroker DLJ's turnover is done via i-mode.

Every i-mode customer uses his or her mobile telephone number with the extension @docomo.ne.jp for all email addresses. All i-mode users together send more than 2 million emails per day. DoCoMo works together with the Internet search engines Excite, Yahoo!, and Lycos. Together with Microsoft, DoCoMo has founded Mobimagic, which offers access to corporate databases, corporate calendars, and email via the mobile phone and the PC at home. Softbank supplies intranet applications similar to Compa, Puma, and NTT data. In cooperation with Sun Microsystems, DoCoMo introduced Appli, a Java execution environment offering information, transaction,

Figure 5.6
Example of a cartoon on i-mode.

entertainment, and database sites. Games can be downloaded and maps can be viewed (see Figure 5.7).

By Japanese standards, the prices are relatively low. The user pays a monthly charge of 300 yen (100 yen is about $1). In addition, the user pays for the amount of data sent, and not, as with the current WAP services, for the time the service is used. A message of 250 characters can be sent for about 4 yen and received for 2 yen. A 20-character message costs 0.9 yen for sending and the same amount for receiving. Most

Figure 5.7
A popular Sony i-mode terminal, the SO503.

services are priced between 100 and 300 yen. For example, Disney pictures cost 200 yen, a Pacman game 300 yen, fortune telling 150 yen, and Formula Nippon info 300 yen. After a customer subscribes to i-mode, his or her mobile voice traffic increases by 15 to 20 percent on average. In Japan, DoCoMo expects that in 2003, the penetration of mobile Internet will amount to 80 percent of mobile users.

Other Developments in Asia

In other Asian countries, WAP has been introduced as well. In Hong Kong, several hundred thousand customers use WAP. Singapore-based Insead Business School loans students WAP phones, enabling them to follow online educational courses. The biggest mobile market is currently China, with more than 70 million mobile users at the end of January 2001, and adding 2 million new customers each month. It has passed the United States as the largest mobile market. China is a standards battlefield as well. GSM and CDMA (see Chapter 1) are competing to be the dominant standards. Currently NTT DoCoMo plans to roll out i-mode in China, Vodafone bought a small share in China Mobile, and Japanese and Korean companies are conquering the dominant GSM players. According to Chinese analysts, the Chinese market is very similar to the Japanese market and they predict a bright future for i-mode services. Mobile Internet in Asia can benefit from the low fixed telephone and PC penetration. Combined with the attractiveness of new gadgets, high levels of competition to drive down tariffs, and a growing number of compelling applications, mobile Internet is expected to gain momentum.

Mobile Internet in the United States

As opposed to what the media states, the European headstart on the United States is questionable. As a result of the various mobile standards, some say mobile Internet will only be possible on the launch of 3G. With the various systems and the "calling party pays" system, in which the owner of the mobile telephone pays for receiving the call, the penetration of mobile telephony is much lower than in Europe and Japan. However, things are changing in the United States. Openwave and Motorola's prominent presence as founders of the WAP Forum indicates that U.S. companies are aware of what is happening in the market. The W/O portal by Motorola, to which 70 renowned American Web sites are linked, is another indicator. The use of email and Internet by large groups of the population is higher than in Europe. Internet penetration, at almost 60 percent, is higher than the European average. PDAs are much more common than in Europe. With the Palm VII, email can be retrieved and Web clipping in the United States is supported. Operators like Nextel offer data and voice services to the business-to-business market via a substantial packet-switched network. About 10 percent of Sprint's current subscribers use their wireless Web service and Sprint expects that to increase to 60 percent by 2005.

Cap Gemini America has conducted research that indicates that at this moment, only 3 percent of American mobile users use mobile data. Of the 33 percent of business users, 11 percent retrieve email via the mobile phone. About 70 percent of the current mobile data users indicate daily use. Cap concludes that the penetration will increase from 3 percent to at least 78 percent within 12 months! As the most important reason, 47 percent of the subjects indicated the fact that the employer will pay for the use of the mobile phone. The massive migration from paging services to mobile data should explain this enormous growth. From all future users, 52 percent indicate they will use email, business information, and personal information, 24 percent will use email and personal data, and 13 percent will use only email. The mobile connection to business applications seems to be the critical factor for the breakthrough of mobile Internet in the United States.

DIFFERENT ROLES WITH MOBILE INTERNET..........

The roles with mobile Internet form a combination of roles for mobile communication and for the Internet. The user has a device and needs a mobile network service to get mobile access to the Internet. The user will select a portal to easily access the information and services needed, which are supplied by providers and merchants that try to obtain the necessary advertising. Hardware suppliers and network suppliers will continue to play an important role on the Internet side of mobile Internet. Who will take on what role and what conflicts can we expect?

Mobile Phone Suppliers

For the producers of mobile terminals, good tidings will arrive again. In January 2001, there were more than 500 million mobile users worldwide. According to Nokia, this number will increase to more than 1 billion users by 2003. The suppliers of terminals are critical in the value chain. Whereas PCs all look similar and brand is not relevant to many potential buyers, handset supplier brands create more value. A customer is not really looking for a connection to a network but for a specific type or brand. The terminal is more than just consumer electronics: It is something personal like a pen or watch. Terminal suppliers are the critical factors in the introduction of new services. This also goes for WAP. For some, WAP is an acronym for Where Are the Phones? After the introduction of the Nokia 7110 in 1999, it was a long time before other phones followed. By the middle of 2001, few phones without a WAP browser were sold. The availability of a WAP browser is a necessity, but no guarantee for success, as proven in the past by SMS.

WAP offers the terminal supplier the following advantages:

- Carrier independence of WAP provides the opportunity to use the same WAP software in different types of devices for different network standards.
- Economies of scale.
- Design of WAP is aimed at minimum memory consumption so that cheap components can be used.
- The knowledge of mobile telephony for the development of integrated Internet telephony services can be exploited.
- Possibilities to introduce new innovative products in order to differentiate brands.

The device that is used to gain access to the mobile Internet can be viewed from different angles. According to mobile terminal manufacturers, it is a telephone with an added browser. According to the PDA and handheld computer suppliers, it is a handheld computer with an added mobile network card, or one that can be linked to a mobile network with a mobile phone. Other applications are also possible, for example, the connection to a refrigerator, car radio, or Walkman to make these devices more intelligent. Which suppliers have the greatest opportunities in this market? The classic mobile phone suppliers have the best opportunities to successfully supply Internet terminals to the market. They are used to dealing with complex dynamic developments of mobile phones from demanding customers. Because of their involvement in the development of mobile networks, they are the first to supply devices that use the new technologies. Palmtop suppliers will be very successful niche markets with mobile Internet terminals, but will have to build mobile telephony into their products at lower costs to be successful on the mass market. Mass-market products have to be priced below $150. There is reason to assume that these manufacturers will want to become active on the market for mobile Internet. In 1998, 1.4 million PDAs were sold in Europe, as opposed to 90 million mobile phones.

As indicated in Table 5.1, the market for smart phones will jump forward.

Table 5.1
Market for Smart Phones

	Western Europe (millions in 2004)	United States (millions in 2004)
PCs with mobile connection	14.5	16.9
Handheld PC with mobile	3.4	4.0
Smart phone	60.8	34.4

Infrared routing and Bluetooth make building a mobile phone in a PDA unnecessary. The PDA can use a mobile modem via Bluetooth to contact the network. The largest PDA suppliers, Palm and Psion, are pioneers. Maybe the addition of mobile is the key to success in the mass market Internet for PDA suppliers. Handheld computers are more lightweight and compact than ever, while still maintaining all important features. Without a doubt, the manufacturers of handheld computers want to capture the high end of the Internet terminal market. The founding of Symbian proves this.

THE FIGHT FOR MARKET SHARE FOR HYBRID DEVICES

According to market research company IDC, the market for handheld computers in the United States amounts to $674 million. According to Strategies, this market is estimated to grow to $2 billion in three years, and in 2005, it is expected to reach $7.8 billion. Market leader Palm has 75 percent of the U.S. market. The Palm VII uses a gateway server. In this way, Palm has introduced a closed mobile portal. Palm aims to introduce this portal in Europe as well. Unfortunately, this is not a worldwide technology and the application is limited to Palm OS devices and cannot be used for calling. For the Palm V, a WAP browser is available (Figure 5.8) so that with the use of a mobile phone, WAP services can be accessed via the Palm. Palm has selected a new processor to continue to support all existing applications for its operating system, and to allow GSM manufacturers to use the Palm operating system. Palm also developed an e-payment system in cooperation with HP daughter Verifone, Visa, and Ingenico.

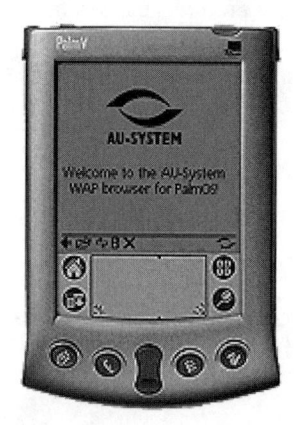

Figure 5.8
Palm V with WAP browser.

To be a player in the market for hybrid devices, or to quote Nokia's CEO Ollila, "to put the Internet in every pocket," Symbian was founded by Ericsson, Nokia, Motorola, and Psion in 1998. Later, a large number of other parties in the communications industry joined this foursome. Symbian uses EPOC, an open operating system for PDAs. The Psion Revo and the Ericsson R380 are equipped with EPOC. EPOC gives the mobile telephones PDA functions and vice versa. The open Symbian alliance did not consider using Windows CE or Palm OS. Symbian used EPOC, but developed applications can also be used with other operating systems. Symbian has issued guidelines under the name Quarts for the development of hybrid devices that support both mobile telephony and wireless computer communication. Symbian version 6 supports HTML, Java, email protocols, and several office applications. It also encourages third parties to develop applications. Microsoft and Palm can benefit from this, as the guidelines are not platform-specific. Nokia already owns a license to the Palm operating system and might use the Palm operating system for an EPOC-running mobile device. The current model, the Communicator 9210, is based on Symbian 6. This enables users to use applications developed by third parties like route planners based on positioning.

Besides Symbian, Palm has competition from Handspring. Handspring has introduced the Visor (Figure 5.9), a smart-looking Palm clone sold at lower prices. Handspring, owned by two cofounders of Palm, uses the Palm operating system and offers "hot swappable" memory modules. With this, functions such as MP3 or a digital camera can be added at any time. Both Handspring and the Symbian-related companies seem to prefer to use the Palm operating system.

Figure 5.9
Handspring Visor with camera.

Another issue is competition from Microsoft. Microsoft introduced Pocket PC to replace Windows CE in the beginning of 2000. Compaq, Casio, HP, and Symbol were among the first to produce the Pocket PC. Pocket PCs offer audio capabilities and options to store e-books, as well as calendar and address book functions. Wireless Internet access and reading emails are possible using a mobile phone, and the mobile phone is integrated in the next models. Microsoft initially dismissed WAP as an inferior form of HTML, but is now a member of the WAP Forum and cooperates with WAP pioneer Ericsson in combining Windows 2000 and Exchange with Ericsson infrastructure. Telstra (Australia), T-Mobil (Germany), and Vodafone (UK) are the first operators implementing Microsoft's new tools that connect data stored in operator and corporate servers to mobile networks. Microsoft developed Mobile Explorer, a microbrowser for HTML and WAP. The company is also working on SIM cards with a new Windows operating system. It entered into an alliance with AT&T and British Telecom to develop software for the business-to-business and consumer mobile phone markets. The applications would be suitable for Microsoft's Pocket PC, but also for EPOC. To service the market together with the PDAs, price is a critical factor. Applications need to be designed from a mobile phone user's perspective and linked to what the PC user knows. Microsoft seems to have learned from the failure of Windows CE, but it remains to be seen if it can follow the simplicity of applications and the price level of Palm, Handspring, and the Symbian partners. The simplicity in integration of communication modules adds an extra dimension. WAP can coexist with every operating system—EPOC, Palm OS, Windows CE, FIEXOS, OS/9, Java OS, and so forth. It is expected that a WAP browser will be added to Pocket PC.

Browser Suppliers

A browser has to be installed on the mobile terminal to access the Internet. Suppliers will come from two areas: The mobile telephone suppliers will install their own browsers on their devices or use the browsers built by software companies. Table 5.2 shows the browser suppliers for a number of WAP device types.

Table 5.2
Browser Suppliers for WAP Devices

Handset Manufacturer	Browser Supplier
Alcatel	Openwave
Benofon	Microsoft
Ericsson	Ericsson
Nokia	Nokia

Table 5.2
Browser Suppliers for WAP Devices (Continued)

Handset Manufacturer	Browser Supplier
Motorola	Openwave
Palm V	Ausystem, WAPman
Samsung	Openwave
Siemens	Openwave
Sony	Microsoft

At the moment, WAP browser licenses are given to device manufacturers (Figure 5.10) that have not developed their own browser. Because the largest manufacturers of both devices and gateways have developed their own browsers, the WAP browser market itself might not be profitable. The development of the browser will then be financed by hardware sales.

Figure 5.10
Examples of WAP-enabled phones from Ericsson, Motorola, Samsung, and Nokia.

Software Tools Suppliers

Just as with the Internet, there are numerous ways to facilitate WAP services. Think of the existing tools supplied by Frontpage or Verisign. The knowledge gained can be used to develop tools specifically for mobile Internet. The suppliers can:

- Reuse existing products for WAP applications
- Find new customers in the mobile world

New parties also enter the market aimed specifically at supporting mobile Internet applications. For example, a French company called Webraska is the pioneer in the area of location destination and graphical applications. Every day, new or existing companies start developing products for mobile Internet.

Gateway Suppliers

A WAP gateway is needed to become an ISP that supplies WAP services. Various suppliers supply these gateways. The most well-known are CMG, Ericsson, Nokia, and Phone.com.

To see how global the WAP developments are and which parties are ahead in the market, refer to Table 5.3.

Table 5.3
Global Status of WAP Gateway Installations

Country	Mobile Operator	Gateway Supplier
Australia	Optus	CMG
	Telstra	Openwave
France	Itineris	Nokia
	SFR	Alcatel
Finland	Radiolinja	Nokia
	Sonera	Nokia/Apion
Germany	Mannesman D2	Siemens/Openwave
	E-plus	Oracle
Hong Kong	Smartone	Ericsson
	Sunday	Openwave

Table 5.3
Global Status of WAP Gateway Installations (Continued)

Country	Mobile Operator	Gateway Supplier
Japan	IDO	Openwave
	DDI	Openwave
United States	Bell Atlantic Mobile	Motorola
	Bellsouth	Motorola, Nokia, Openwave
	US West	Openwave

Even though WAP is the de facto standard, not every WAP browser will work well with every WAP gateway. The WAP gateways by Nokia and Ericsson, for example, only support the Openwave browser with version 4.1 or higher. This could mean that not all devices with Openwave in the Netherlands will be able to use WAP. For the portal, hardware and software are needed to supply portal services to the users via WAP, and also via the Web. For example, Oracle supplies a portal-to-go, with which portals can be created that can be accessed via Web and WAP on all possible devices. The portal offers facilities for database switches, payment, personalization, location determination, and voice recognition. Gateways don't have to be big, complex, and expensive. Gateway suppliers supply their products not only to operators, but also to companies that prefer to use their own gateway. In these vertical markets, the bank sector and the IT sector are especially active. Internet service providers are also important buyers of WAP gateways. Other companies will follow. A UK company called V3 offers a combined WAP/SMS gateway, which can be downloaded from their Web site. It also provides additional applications and APIs to develop your own applications. You can start small for a relatively low price ($400) and add additional features and capacity when you need it.

The gateway suppliers play an important role in the mobile Internet value chain. The speed at which new developments in the gateway are implemented and the scalability, quality, and compatibility with various types and brands of devices and browsers are crucial factors for success for suppliers of mobile Internet services.

Mobile Operators

For the mobile operator, mobile Internet could be a blessing in disguise. Mobile Internet provides the operator with new opportunities:

- Increasing the turnover per customer by generating more traffic

- Giving the customer relationship another dimension with a visual interface that can make those complex services easy to use (i.e., telephone services such as voice mail, call waiting, or three-way calling)
- Differentiation in respect to the fierce competition
- Developing new market segments
- Creating customer loyalty for a longer period of time
- Developing new services in cooperation with various partners and decreasing dependence on the classic telecom suppliers
- Decreasing customer care costs by offering information on their own products via WAP

On the other hand, mobile Internet is a threat to the position of the mobile operator as owner of the customer. How mobile Internet will affect the operator depends on the other roles the mobile operator can create for itself. If the mobile operator is able to become an Internet service provider and portal supplier, it can make the relationship with its users stronger. By collecting data on the use of mobile Internet and the different services, the mobile ISP and portal supplier can learn more about the needs of individual users and how to link these needs to the services provided. If the mobile operator only fulfills the role of data transporter ("data pipe"), and others fulfill the role of mobile ISP and portal supplier, the role of mobile operator could reduce or eliminate contact with the customer.

With this, mobile data traffic threatens to become a low-margin product that can be purchased in bulk by mobile Internet service providers and later offered to the customer in a complete package deal. This could still be very profitable for a mobile operator. With the high prices that have to be paid for 3G licenses and networks and the large investments that mobile operators have already made to develop customer relations, however, the role of data transporter does not appear to be the most profitable for mobile operators.

Which mobile operators have the best chance to come out winners? The mobile operators that entered the market for mobile communications first probably have the best chances in any country. They no longer have the pressure within the organization to meet the basic conditions to play in the mobile communications market with nationwide coverage and a large customer base. They can more easily afford resources (money and people) for the innovations needed for mobile Internet as opposed to operators that are still building their network and customer base.

Mobile operators that are active in many countries and are able to introduce inventions from the native country in other countries have the opportunity to become successful. They can gain a technological competitive advantage if they use the same technologies in all countries, saving costs by buying in volume. They can also obtain better deals and preferential treatment from suppliers. Global content providers can also make better deals, as they have a presence in many countries and have many customers. In reality, obtaining synergy by offering services in many countries is difficult, as markets for mobile communication are not all developed to the same extent. The same content may not be interesting in every country. In the Netherlands, traffic jams are an important aspect of everyday life. In France, the need for traffic information mainly applies to Paris. In the rest of France, it's important to book a motel next to the highways. In the United States, business travelers take planes instead of cars, meaning demand for hotel bookings near airports and flight schedule information.

The mobile operator has a strong position in the mobile Internet market for a number of reasons: knowledge of mobile networks, the SIM card they offer the user, the location information, the invoicing relationship they have with the customer, the distribution strength and its influence on the product line, the strength of communication, and knowledge of the local market. However, they may not appear to be the most successful. Mobile Internet demands a different service development process, a much higher speed of transactions, strong partners, and guts. We cover this in more depth at the end of this chapter, in the section "Conflicts in the Value Chain."

MEMBER PROPOSITION

> KPN Mobile supplies a member proposition with its @info services, with which users who register as members get benefits. These benefits are the use of a Lotus planner, address book, and to-do lists, which can be accessed via the Internet and via WAP. Members can also create their own site layout via the Internet for WAP, so that they can access these frequently used services quickly via WAP.

Mobile Internet Service Provider (M-ISP)

The mobile ISP provides Internet access to mobile users. This means that the mobile ISP has to purchase a WAP gateway that has to be connected to the mobile or fixed network and that can function as a dial-in pool for the Internet. Many mobile operators will try to fulfill this role and find it to be a natural addition to their regular Internet access service. In the Netherlands, XS4ALL was the first Internet service provider to offer these services and therefore the second WAP service provider, after @info. Similar to the Internet, the user has to enter dial-up connections to the browser of the mobile terminal. For many devices, this is difficult. To stress the complexity, we have included a display showing device setup procedures. Some device manufacturers such as Nokia and Ericsson understood this from the start and offered a special mechanism with which setups can be sent to the user via an SMS and accepted and saved easily.

DEVICE SETUP PROCEDURES

> To use WAP services, the browser has to be set up: the home page, the type of connection, the security of the connection, the speed of the connection, the dial-in number, the IP address of the WAP gateway, user name, and password. As an example, we give the instructions that XS4ALL gives to their Nokia 7110 users:
>
> When you have just switched on the phone, press *menu*
>
> Scroll to services, and press the scroll button
>
> Go to set up and then to connection functions
>
> Choose *Set 2* and press *New Name*
>
> Type in *XS4all* and click *OK*
>
> Select the connection *XS4all* and choose *edit*
>
> At Homepage type *http://wap.xs4all.nl/* and press *OK*
>
> Select *Continuous* for Type of selection,
>
> For Selection security, *off* and for Carrier, *Data*
>
> Type *0204043652* for the dial-in number and select *OK*.
>
> For IP address select *194.109.209.131* and select *OK*. Select *Normal* for legalization type.
>
> The Type Datacall has to be ISDN with a mobile data subscription. If you do not have a data subscription, analog. Should ISDN not work, select Analog. Speed data call should be at 9600. Finally, type in the User ID and password at the respective fields.
>
> Consult your phone manual for other questions about your Nokia 7110.

With the Internet, less than 50 percent of the users change their browser setup after installation. We expect that in light of the complexity, this percentage will be even lower for mobile phones. Most devices can store only five different setups. The WAP browser also offers the possibility of bookmarking. The number of bookmarks is limited to 10 or 20. This means that the customer can store up to five mobile ISPs and a maximum of 10 to 20 sites. Elimination of this complexity is a condition for bringing the mass market within reach of the traditional Internet service provider. It is important for the service provider to be added to the different setups the user has at his or her disposal. In Germany, YourWap pays retailers to have their services added to the settings. Siemens installed settings for Yahoo! in some devices. Gartner predicts that a large percentage of the cost of PDAs will be offered for free by Internet service providers, in combination with a long-term contract, in a manner similar to the current mobile phone model.

Mobile Portals

This is the role for which there are the most candidates, all from different camps. The mobile portal supplier supplies the starting page from which the user can access different mobile Internet services. It concerns applications such as planners, email, messenger services, and other relevant information. Notification of appointments and relevant news are also part of the standard portal functionality. With member propositions and registration of the services used, the mobile portal supplier can collect valuable customer profiles. This information can be used to make relationships with the user stronger and to improve the supply of services. Similar to the Internet, a wide variety of generic and specific portals will appear. Who are the most important candidates for this role?

- *The mobile operator.* Most operators offer a mobile Internet portal. Vodafone and Vivendi will offer worldwide Internet services for mobile telephones under the name Vizzavi in all Vodafone networks (see Figure 5.11). They have formed an alliance with leading technology companies such as Sun, IBM, Psion, Palm, Nokia, and Ericsson. Vodafone also announced a joint venture with Cap Gemini Ernst & Young named Terenci, which aims at offering B2B applications for the mobile market, especially timely mobile information for vertical markets like freight and logistics. In the United Kingdom, British Telecom launched Genie a few years ago, offering SMS information services. Integration of WAP has made Genie one of the most successful portals, with more than 1 million active WAP users by March 2001, and hundreds of thousands more each month. British Telecom has rolled out the Genie consumer concept from the United Kingdom to other BT networks. The mobile portal has to grow into a place where information is gathered and services are available at all times. Service providers such as Canal+, Infospace, stockbroker Charles Schwab, and travel agency Travelocity are also part of this alliance. The French version offers telephone services, email, contact lists, calendars, and search functions, together with information and communication services via Internet, SMS, and WAP.

Figure 5.11
The Web home page of Vizzavi, the mobile Internet portal of SFR in France.

- *The Internet service provider.* In February 2000, AOL announced it would set up mobile portals with email, news, and weather reports for Europe in cooperation with Nokia, Ericsson, and RTS Wireless. UUnet (part of WorldCom) and WorldOnline, in cooperation with Ericsson, have also indicated that they will set up mobile portals. Deutsche Telecom has a keen interest in Freeserve, the largest Internet supplier in the United Kingdom. The combination of DTs mobile operator in the UK, One-2-One, and Freeserve offers many opportunities, especially with the One-2-One-obtained UMTS license.

- *Internet portal suppliers.* In Germany and Sweden, Yahoo! together with mobile operators offers an address book, email, financial news, weather, and My Yahoo! to create a personal news environment. In the Netherlands, ALLWAP.nl supplies information via the Internet and via WAP links via *mmm.allwap.nl*. MSN and Yahoo! have portals in the United States that are suitable for types of mobile terminals and networks that cannot be used in Europe. The so-called headstart the European companies had over U.S.-based companies is very questionable. The large American portals have created their own mobile divisions and work together with European mobile operators. Excite works closely with the Japanese DoCoMo. The lag of the Americans in the area of mobile services is compensated for by the strength of their brands and their considerable knowledge of the Internet.

As mentioned earlier, the diversity of mobile telephone standards might delay mass introduction of mobile Internet to the American market. As a result, many smaller American Internet companies may decide to focus on the regular Internet instead of the mobile Internet. However, the initiative of WAP came from (at that point in time) a small U.S. company, Palm (Figure 5.12).

- *Content providers and merchants.* Amazon.com, Dutch Bruna, and Scandinavian Boxman already supply books and CDs via WAP. Many content providers and merchants will try to create a portal based on the specific interests of their target audience or by offering a wide selection of links. Time Warner has a whole range of content at its disposal, from CNN to cartoons, and it now has the knowledge to add Internet access to this content, as the result of its merger with AOL.
- *Transaction suppliers, such as banks.* Banks could fulfill an important portal function if they can offer payments, investments, and other financial matters via WAP. See the sidebar, "Banks and Mobile Payments," later in this chapter.
- *Device suppliers, browser suppliers, and/or gateway suppliers.* For the development of devices that are suitable for WAP, many device suppliers wonder if services will be available when devices are introduced. Ericsson has created a demo portal to show the options and a joint venture with AOL. Various suppliers have signed contracts with content providers so that they themselves can supply WAP services in case the mobile operator

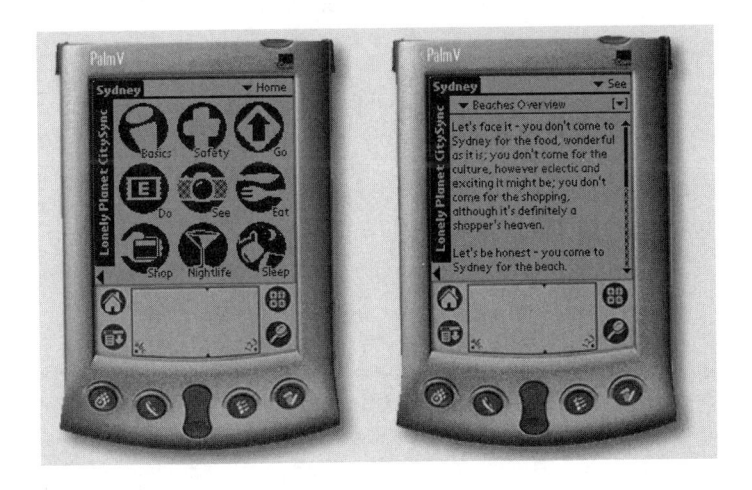

Figure 5.12
Example of the Lonely Planet Travel Guide on the Palm V.

does not. Nokia has an agreement with CNN, but also with several French Internet startups. Wireless Entertainment Services offers all Nokia customers personalized dial tones. Nokia also has initiatives like Club Nokia, which sells downloadable ringing tones based on EMI's catalog. Lucent has created a portal with Zingo, a Siemens joint venture with Yahoo!. Motorola has launched a mobile Internet Exchange to which more than 70 large Internet sites are connected. Amazon.com, Reuters, Sports.com, and Worldspan, one of the largest online reservation systems, are some of the other players. Operators have already forced Ericsson and Siemens to retreat from their portal ventures.

- *Strong brands.* A recent interbrand survey put Nokia as the fifth most recognized brand in the world, ahead of any telecom operator. Strong brands, such as Virgin, Disney, or AOL could play an important role in this market as virtual network operators with the support of technology companies. Other nontelecom companies with a strong brand and a focus on specific user segments can provide these users with the services they want.
- *New companies.* Examples are 123Internet (see the case in Chapter 8) and Wapportal. Wapportal is an independent Danish company that offers services via WAP and Internet to Denmark, Norway, and France. The company gets its revenue from participating content providers and not (yet) from users. E-commerce and B2B services are added, as well as services that give content providers online customer profiles. The intention is a worldwide rollout in cooperation with local operators. Along with generic and personal information, the portal also offers country-specific links. Users can access information on their native country while abroad. Another example is a UK portal named BoltBlue that claims to be the UK market leader in terms of registered users. The portal offers an interactive fixed Internet site with SMS services and a WAP site. BoltBlue claims its success is purely based on word-of-mouth recommendations. The portal, with services including news, sports, mobile games, local travel, and weather for some UK areas like London, is aimed at people between 17 and 30 years old. BoltBlue makes most of its revenue through advertising on the Web and WAP.

Many parties have a good chance of success. Forming alliances to obtain missing expertise is an indication of future competition. However, there will be more winners: the market leaders with the winning product-market combination for a specific market segment. Later in this chapter, the factors for success are discussed in greater detail.

Transaction Providers

The role of transaction provider can be fulfilled by banks and credit card companies, but also by specialized mediators that can maintain the relationship with banks and credit card companies. The partnership between transaction providers and mobile operators is very strong. The mobile operator could then provide extra security measures and payment applications, thanks to the distribution of the SIM card, while the transaction provider finalizes the actual payment of the transaction. Executing secure transactions is something many parties would like to take ownership of. With the widely expected explosion of the number of mobile transactions, successful transaction standards appear to have a golden future. Finnish operator Sonera has founded a consortium called Radicchio in cooperation with 35 other companies. Radicchio developed a standard for electronic trade based on a finding by Sonera.

Nokia has been working together with the Finnish bank Merita-Nordbanken on a payment standard for some time now. Ericsson is working with Visa International on a safe method of payment for the (mobile) Internet, with a focus on Smart Cards. Ericsson has mentioned a Wireless Wallet. This is a mobile phone with Bluetooth and WAP options that will make credit and debit cards unnecessary. Nokia, Ericsson, and Motorola have joined forces and developed a similar initiative in which other suppliers can participate. This initiative should result in a so-called personal trust device, which adds extra security to WAP. Suppliers of Smart Cards and banks also participate in this race. Banks are often looking to cooperate with telecom operators. In Spain, Banco Bilbao Vizcaya Argentaria and La Caixa have a strategic stake in the largest Spanish operator, Telefonica. Together these companies bought an Irish Internet bank.

BANKS AND MOBILE PAYMENTS

> Payments stem from the intention to buy or donate a product or service. The relationship between buyer and seller is much more important to the payee than the relationship with the bank or the method of payment. Payments will sooner be a qualifier for mobile Internet than a unique selling proposition. Absence or imperfection of methods of payment lead to dissatisfaction, but the presence of effective methods does not lead to increased satisfaction. Payments have to be executed cheap, fast, easily, accurately, and free of risk. The techniques under development have contributed to these objectives. The role of banks has always been a dominant one, as banks had access to the knowledge and the means to execute safe payments on a relatively exclusive basis. This knowledge is currently also collected by nonbanks. The role of banks

is slowly being reduced to that of supplier of wholesale building blocks such as electronic banking or back-end transaction handlers.

Where the banks previously made their own payment products, the payment products of the future will only contain the bank's payment engine. Nonbanks (telecom operators or Internet portals) purchase these products and repackage them for sales. Outside of the mobile Internet sector, this continues. Easypay by Shell is a product largely developed by Shell with which it wins back authority over the customer with regard to the payment process at the gas station. In light of mobile Internet, it goes without saying that banks will form alliances with ICT companies. Founding the Mobey Forum is a good example of an alliance between banks and technology companies. Mobey aims to integrate financial services with mobile telephony. Larger banks such as the Dutch ABN Amro, the Spanish BSCH, the French BNP Paribas, Deutsche Bank, and Visa work together with Nokia, Ericsson, and Motorola. The development of services is based on the previously mentioned development of the personal trust device by Nokia, Ericsson, and Motorola.

Content Providers

Mobile Internet offers content providers a number of options:

- Reduction of cost
- M-care, which is customer service by supplying information services, resulting in margin improvements
- M-billing, or invoicing and sending reminders via the mobile phone, resulting in margin improvements
- Presence, which makes the supplier available to the customer at all times, regardless of hours and physical locations
- M-commerce, which is generating turnover from existing or new customers via the mobile phone

Similar to the start of the Internet, the parties are divided. On one side are those who wait to see what will happen, and on the other side are those who want to gain experience right away and start transferring current services to WAP services. Companies with large information databases such as Yellow Pages, telephone directories, and Scoot have seen the added value and already offer their services via WAP. Suppliers of news services, such as newspapers, have not been waiting as long with WAP as they did with the Internet, maybe because WAP allows them to calculate a fee for their services.

Reuters distributes its information via various channels: via its own portal in a number of markets, via partnerships with Ericsson and Nokia, and via existing portal sites such as Excite and Yahoo!, which are also building a mobile portal. We expect that in the end, the content providers which are able to supply detailed current information related to the personal preferences and the location of the user, will add the most value. These could be new parties that get ahead of the competition by insight into the market. Without a doubt there will be mobile Amazon.coms that get a headstart in a particular market. We believe that in the past, the existing player in the market received a wake-up call with the introduction of the Internet, and this time around they do not want to play catch-up again. The interest from large companies in presentations on mobile Internet proves this. Existing parties in the market can use their physical locations and their well-known brands to improve services for their customers.

Datamonitor expects that the global mobile content market will generate $32 billion in 2005, compared with $2.4 billion in 2001. Datamonitor expects entertainment to be the largest revenue generator in mobile content, with $11 billion in 2005, followed by commerce, multimedia, and location-based services.

Merchants

The Internet merchant, such as book, CD, and computer equipment suppliers, have been the first to also offer their services via WAP. Mobile access to their services provides them with the advantage of impulse buys. The mobile relevance of their services so far has been very limited. We believe that the following types of merchants will have an excellent shot at success:

- *Vending machine distributors.* If you can pay via the mobile phone, you do not have to forgo a purchase just because you do not have small change. Parking services can also be improved by allowing reservations and payment via mobile phone.
- *Last-minute ticketing.* Airline, train, and theater tickets can be sold faster to people who previously indicated an interest in these tickets.
- *Game and music suppliers.* If you can download music and games and play them directly on your mobile terminal, this offers a greater advantage than the Internet. Personalized radio is one such option, whenever and wherever you want it.

Even more important than with the Internet is the transaction process of the merchant. When you purchase something via mobile phone, for example, a movie ticket that you want to use within 10 minutes, delivery has to be imminent. For services such as ticketing this can be done by issuing a password that the customer can use to receive the ticket. In the case of vending machines, the product purchased needs be available right away. If it is not, the customer is likely never to purchase from that source again.

Advertisers

A mobile device is a very personal device. Advertising can therefore also have a big impact, as you can communicate one-on-one with your target market. Location determination can make advertising very targeted. As soon as the advertisement is not aimed specifically at the user and the situation, the advertisement will be irritating, especially because the user will often pay for connecting, downloading, and viewing the advertisement per minute or per byte. For advertisers it is therefore necessary to be relevant and to the point and to come up with interesting trade-offs. In return for the opportunity to advertise, the user could get extra call time or free information subscriptions. U.K.'s BoltBlue says advertisements need to be positioned very strategically and should be related to the content the user chose to see and the user profile. Sky Go conducted a trial with a thousand users in Colorado receiving at least three wireless adds per day. Statistics were amazing:

- 64% of the ads were opened.
- 58% of the branded ads were recalled.
- 15% of the 58% resulted in action or planned action.
- 2.9% of the ads resulted in online or offline purchases.
- Only 0.7% of the click-to-call ads were activated.
- 37% of the users asked for credit card information during a transaction provided it.

Although the novelty of the medium has a lot of impact, these statistics are positive, if you compare this 58% recall rate with the 0.04% for banner ads or 15% for streaming media.

Unwanted advertising via SMS (SMS spam) has an even higher rate of irritation than advertising via email. Imagine that your mobile phone beeps every five minutes for some unwanted message. The Amsterdam Police use this irritation in another way: They send SMS "bombs" to mobile phones reported stolen. The thief receives a message every few minutes, displaying a warning, in an attempt to reduce mobile phone theft. SMS spam has caused a lot of irritation, with users blaming operators for unwanted SMS. DoCoMo has had similar problems with unwanted emails sent to its subscribers' mobile email addresses.

Similar to Internet advertising, the main advantage of mobile advertising is that the results of a campaign, such as reach and response to the advertisement, can be measured more accurately than with traditional media.

All parties involved in the mobile Internet industry have to be cautious that the advertising market is not ruined and that users will no longer use the medium for advertising. Just look at the effects of telemarketing and commercial television channels, with commercial blocks in movies. The Finnish telecommunications company Sonera has made a pact with Icon Medialab and the advertising agency 24/7 to develop joint marketing systems for mobile appliances. The advertisements are presented to users via SMS and WAP. In the Netherlands, the Marketing Association has founded a mobile marketing/mobile commerce workgroup that will research issues such as permission for sending advertisements to the mobile phone and developing interaction models for WAP advertising. Table 5.4 gives an overview of the roles in the mobile Internet value chain.

Table 5.4
Overview

Roles in Value Chain	Explanation	Examples of Suppliers	Paying Party
Content provider, merchant	The content provider offers information, communication, and entertainment services. The merchant sells products and services.	Disney, CNN, Weather.com, Citibank, Bloomberg, with, respectively, pictures, news, weather information, banking services, and financial news Banks ISPs such as AOL and Compuserve with their own content	End user (on account of the telecom operator or other methods of payment); advertisers
Hosting provider (Web/portal hosting)	Content runs on a WML server to which database applications are added	Operators, ISPs and ASPs, content providers Portal-to-go by Oracle and Cap Gemini	Information supplier

Table 5.4
Overview (Continued)

Roles in Value Chain	Explanation	Examples of Suppliers	Paying Party
Access provider for information supplier	Parties that provide content for the Internet; these parties could have a WAP proxy	T-online, UU-Net, Worldonline, Compuserve	Information supplier
Transaction provider	The party that handles the financial transactions	Banks, trusted third party, telecom operator	Information supplier, advertisers, and maybe the end user
Portal	Mobile Internet portal	i-mode (NTT DoCoMo), Vizzavi (Vodafone Airtouch), Genie (BT), Excite, Yahoo!, Interspace, Startwap, ISPs	Information supplier, advertiser, and maybe the end user
Network operator	Telecom network operators develop and maintain the network and suppliers such as Nokia and Ericsson supply centers to the operators for this purpose	AT&T Wireless, NTT DoCoMo, Orange, KPN Mobile, Vodafone Airtouch	Mobile service provider
Mobile service provider	Party that provides access for the user, by mobile subscription with a help desk; the customer is invoiced	Operator or virtual operators like Virgin, Talkline, Debitel	End user
End users with device	WAP device (including browser) or PDA with WAP browser and mobile device	Nokia, Ericsson, Alcatel, Motorola, Palm, NEC	End user subsidized by the network operator or not; service provider

CONFLICTS IN THE VALUE CHAIN

Even though we already discussed the ambitions of the various parties in the previous section, a clear perspective of the most important conflicts should be presented. To predict which conflicts will arise, it is important to first determine where the most value is held. The following activities supply the user with the most added value:

- Mobile access
- Advertising
- Location-based services
- Transactions
- Customer ownership

In case of a gold rush, it is often smarter to supply the miners with products and services than to look for gold yourself. The chances for success are better. The suppliers of gateways, devices, and specific software, and of course the consultants, might very well be the winners.

Mobile Access

The first obvious link to success is mobile traffic itself. Even though the margin on mobile traffic is not as large as in the past due to increased competition, it is still an interesting market due to the growing user group. With the introduction of mobile portals, information providers, merchants, and transaction providers to the mobile network, the competition for the mobile traffic dollar will intensify. These parties will reason that, as a result of their customers using their services, extra minutes or megabytes are generated, from which they should benefit. The mobile operator will reason the opposite, as they are the mediator and bring the customers in contact with the mobile portal, the merchant, the information provider, and the transaction provider. Therefore, the operator should be rewarded. Operators follow different strategies: NTT DoCoMo and KPN Mobile have revenue-sharing offers with third parties, whereas others like Vizzavi buy content from providers. In Europe, many companies are looking for opportunities to sell airtime as well, using the radio network of a mobile operator. They aim to become a so-called mobile virtual network operator (MVNO). National telecom authorities encourage MVNOs because they stimulate competition and will bring down prices. Mobile operators have to decide if they want to use their radio network to gain additional revenue by launching a new competitor. This new competitor might decrease the operator's existing revenue by capturing the operator's market share and bringing down retail prices. The price structure of the MVNO deal outlines the freedom an MVNO has in determining its own tariffs. There's a big difference in a cost price plus price structure and a retail minus structure. Whereas MVNOs in 2G networks are rare, expectations are that MVNOs will be important players in 3G networks, addressing specific customer groups with targeted offerings.

Advertising

Selling advertising space will be another important source of revenue. Independent UK mobile portal BoltBlue makes the bulk of its revenue through advertising. Revenue-sharing deals (of SMS messages) deliver a minor contribution. If a mobile portal or information provider has many visitors, it's attractive for advertisers. The more knowledge a portal has about visitors, the more revenue targeted advertisements will bring. The decreasing revenues of Web advertising on the fixed Internet raise concerns about expectations of mobile Internet advertising.

Location-Based Services

Operators expect location information to be an important revenue stream. Combining a personal profile with the location of the user offers services a big advantage. There is, however, concern that third-party advertisements pushed to mobile phone users might be illegal in some countries. The UK's Data Protection Commission warned of the possible illegality of third-party location-based advertising. This means that the location information of the subscriber must stay in the network, offering the mobile operator the exclusive opportunity to offer location-based services. This might be a big disadvantage for Internet companies moving from the fixed PC to the mobile environment.

Transactions

Transactions can be a very important source of revenue for different parties. The merchant profits from transactions, as it can sell its products or services, probably at lower cost than in a physical store. The mobile portals and information providers will want to profit from this by asking for a percentage of the transaction turnover as a trade-off for a referral to the merchant from their site (comparable to paying for shelf space in the supermarket). The transaction provider will also want to receive a percentage of the transaction in return for processing payment. By developing a transaction system, Finland-based mobile operator Sonera wants to grab a piece of the pie. Their solution is being tested by Pizza Hut in Helsinki. The customer calls a special number and payment is deducted from their bank account or charged to their credit card.

Customer Ownership

A large customer database combined with a lot of customer information seems to be most important. That is why a large battle will take place in the area of mobile portals. This role is seen as customer ownership, even more so than with information providers or merchants. The party with customer ownership has the largest margins, it can up- and cross-sell as a result of its customer knowledge, and it can serve new customers with similar profiles. In the early days of the Internet, selling a customer base was a very profitable business. The same will certainly happen as soon as mobile commerce becomes important for every company. The French Cartel authorities have already indicated that operators cannot claim exclusivity in the relationship with the mobile surfer.

"WALLED GARDEN" AND WAP LOCK

> Mobile portals seek ways to let their customers always use their own portal. Sometimes they are paid by information providers, who pay for space in the menu structure of the portal or for a more prominent slot in the search results. The tendency arises to disable the competition by limiting the freedom of the user. This can be done by only giving the customer access to information of the information suppliers associated with the portal, the so-called walled garden concept, a garden in which the user can move around freely, but is fenced in. The customer can therefore not visit every WAP site on the Internet. History has taught us that users do not accept this. AOL started a battle with Vodafone in the beginning of 2000 against the introduction of the walled garden concept.
>
> Another possibility for limiting the freedom of the customer is the installation of a WAP lock in the browser of the device. The user can then access information only via the portal. The user cannot choose another portal or circumvent the portal to get to a WAP site. A supplier can only do this in cooperation with a device supplier. In May 2000, France Telecom was sanctioned by the French Cartel authority. Offering WAP suitable devices with such a WAP lock was punishable with an $80 fine per offense. In the United Kingdom, BT's daughter company Cellnet received a warning for offering mobile phones with WAP lock. Orange already provides a WAP lock, as does Sonera in Finland.
>
> The concepts of walled garden and WAP lock are probably of interest to companies that want to use them, so that employees can always access the intranet sites via WAP, and surfing options are limited to those

> sites needed for work. In commercial propositions to the consumer these options are not as valuable. Operators with a considerable market share probably have the TFC knocking on their door. Consumers will be disappointed in the service when they notice their freedom is limited. The consumer wants to determine which sites he or she wants to visit and which portal will be used.

The mobile operator has customer ownership for mobile voice. This means that because the mobile operator supplies a SIM card, sends the bill, advises the mobile device, and answers customer questions, the customer most often has a relationship with the mobile operator and not with the parties behind the scenes. From this strong position the mobile operator can easily take over the roles of mobile access provider and mobile portal. However, banks, Internet service providers, Internet portals, information providers, and merchants will also want to play these roles. Device suppliers will play an important role in the fight for the mobile portal, as will resellers of mobile terminals. They have a large influence on which setups are preprogrammed on the device. Mobile portals have a vested interest in solid contacts with device suppliers and channels of distribution of mobile phones.

Brands

The importance of branding will become obvious. Brands such as Yahoo! (Figure 5.13) can supply the customer with a mobile phone and determine the home page, and thus customer behavior at the same time. Yahoo! functions as a mediator for BBC news, Reuters, and Associated Press. In the meantime, Virgin has started with mobile telephony in the United Kingdom. Pioneers in the field of mobile Internet will also have to be able to build a brand, just like Amazon.com and Yahoo! did on the regular Internet. Operators are scared big brands will take over end-user control. Nokia is one of the companies worrying the big operators. Where they could intimidate Ericsson and Siemens, Nokia seems to continue its strategy of leveraging its own brand to launch end-user services. The expansion of Club Nokia into a mobile portal is direct competition for the big operators building their own portals. Smaller operators might be happy with Nokia's initiative because it increases usage of their networks in a way they are not able to do themselves. Nokia's brand is so strong that operators cannot afford to stop selling Nokia phones. The only thing they can do is diversify their handset base and push other handset brands.

YAHOO!

Figure 5.13
Yahoo! A strong Internet brand.

With all the millions involved in this growing market, you almost forget where the power really lies: with the consumer. The consumers determine if the services offered really add value to their way of living and working. This means that the added value to the customer has to become the point of focus. You have to be better than the competition. One thing is for sure: The consumer has wide selection and the transparency of the market is closely guarded by telecom authorities. The consumer can choose which services he or she wants to use and from which supplier he or she wants to purchase them. The consumer can also easily join others to obtain lower prices. To determine which added value mobile Internet could offer to the users, we cover the trends in society and the role of the mobile Internet in these trends in Chapter 6.

6 Mobile Internet: Get on Board Now, or Wait?

In this chapter...

- Trends in Society   151
- Start Now or Wait   166
- Mobile Internet in Business-to-Business Relationships   168
- Mobile Internet as Trend Accelerator   169

"Everything you want to know is available on your mobile phone" read the promise by KPN Mobile on the introduction of @info. Internet via the pocket phone is within reach for a large part of the population that will use a wide variety of services en masse. Services that, like the Internet claims, are made by everyone for everyone. Services that offer information, enable transactions, and execute payments. Services that can be used easily and that are tuned into individual needs. Services that can be used everywhere and at all times: at home, at work, in stores, and abroad. Services that use Internet technology can be developed easily and at lower cost. Services that supplement the Internet service expand the Internet experience. In short, there are plenty of opportunities for entrepreneurs who want to offer mobile users new possibilities.

In previous chapters, we explained what WAP is and where WAP stands in relation to the developments in mobile telephony. Also, the differences and similarities between WAP and the Internet were explained. We indicated that WAP is the first version of mobile Internet and why the rapid growth of mobile telephony and the rapid growth of the Internet are indicators for the success of mobile Internet. This chapter answers the question of whether mobile Internet will be a standalone technology push, or if it can be related to the other changes in society. Put in a different way, will people be hot for mobile Internet? The question of whether suppliers and organizations can ignore the changes or have to accept them will also be answered.

A development over time is called a *trend*. Trends pertain to thoughts and behavior with regard to a great number of things. Mobile etiquette is an example of changes in thoughts and behavior that have taken place in a short time span. A few years ago, using a mobile phone outside of the car was considered "showing off." Using the mobile phone while shopping or in a bar is now considered normal. However, it is considered inappropriate to answer a mobile phone during movies, church services, or dinners. The speed at which these thoughts and behaviors change varies. Trends in politics and religion change at a much slower pace than the way in which people change their households.

With this trend, the adoption of products and services can be indicated. A trend that arises slowly becomes more powerful because it spreads and takes up a longer time span. A trend differs from a fad. A fad becomes successful in a short period of time and then disappears completely. Think of Rubik's Cube, the yo-yo, and virtual pets. Then there is the hype. Something is considered hype if it is often discussed in the media. The product does not yet have to be available, or if it is available, it may be hardly used. Sometimes companies try to create hype about a product to create demand or to differentiate them from the competition based on innovations. Good examples of hype are the electronic dog by Sony, Aibo (see Figure 6.1), Playstation 2, new software releases by Microsoft, and the release of new movies. WAP can also be regarded as hype, between the first announcements for WAP at Europe's major ICT

Figure 6.1
The robot dog Aibo by Sony.

event, CeBIT in 1998, and the first commercial introduction a year later. After the introduction, the hype really took shape as a result of the mass communication by all parties involved and the enormous media attention. The service was hardly used, however, due to the scarcity of WAP devices. Now that WAP portals seem to appear out of nowhere, every large manufacturer has a WAP device in its product line, and large companies welcome WAP. It appears to have become a big business, but it created expectations that could not be met. To become successful, mobile Internet must be accepted by the user as an everyday aspect of life. This requires more than pretty phones and media attention. Besides technological improvements, the services offered have to be in sync with the other trends in society. This will determine success or failure of mobile Internet.

TRENDS IN SOCIETY

The rise of the Internet and mobile telephony is a trend by itself. As described in the previous chapter, the use of the Internet has spread rapidly. The mobile phone has also gained in popularity. Mobile Internet, therefore, also has a good chance to grow from a gadget to an instrument that is part of everyday life. To indicate how mobile Internet will penetrate our daily lives, a number of trends are explained. We use the trends described by Faith Popcorn in her book, *Clicking: 17 Trends That Drive Your Business—and Your Life*. The trends are translated to current application possibilities and conditions for mobile telephony, Internet, and mobile Internet.

Downaging

The decrease of importance of age can be explained by the activities that many older people participate in. Their good physical and mental health enables them to play sports, travel, and study. The exchange of time and money also illustrates the increase in the freedom of choice in society. Examples include early retirement, sabbaticals, education, and part-time work. The younger generation will go through the cycle of studying, working, and relaxing after retirement multiple times during their life.

In the last few years, the VCR, video camera, game computer, and answering machine have reached the living room, followed by numerous gadgets with chips and plugs for the kitchen. Then the PC with Internet access and the mobile phone made their entry to the market. Younger people have seen the advantages of these devices and managed to operate them easily. The increasing number of people 55 years of age and older that are online indicates that this trend has spread throughout society. The older population still has to be taken into consideration. With regard to purchasing power, they prove it is worth exploring the possibilities of mobile Internet for the elderly.

There is a difference in the way generations are serviced. In the world of the Internet, three generations are mentioned. The "daily paper" generation was born prior to 1965. These people are well informed and have wide interests. The use of the computer is passive and aimed at text. Decisions are made rationally. This generation reads manuals. The second generation is the "MTV generation," those people born between 1965 and 1985. Their audiovisual orientation is developed strongly and interactivity in thinking is higher than with the daily paper generation. Decisions are often made based on impulses and emotions. The "Nintendo generation" is the youngest generation. These people think interactively. Modern communication is fully integrated with everyday life. This group is able to make decisions rapidly based on short audiovisual impressions. Books and manuals require much concentration. Decisions appear to be impulsive but are very rational.

Mobile Internet can capture the interest of young and old. Besides the fact that mobile Internet offers the possibility to communicate in a way that fits the individual, the age of the user can determine the way information is presented and the preferred media for some services. The situation—for example, work, study, or travel—will determine the need for information. Design and control of the device will also have to be differentiated. Society's interest indicates that there is potential for offering a large variety of services.

Conditions that spring from the decreased importance of age include the following:

- The older generation is not yet out of reach for mobile Internet. The opposite is true. They have the money and the time to delve into the mobile Internet.
- Not only different content or services, but also different service design (e.g., phones) is needed to reach different generations.
- To reach the older population, services have to be adapted to the way this generation treats information (e.g., manuals). The same goes for the younger generations.

Clanning

A clan is a group of individuals with the same goals, objectives, and views. Groups are formed based on mutual interests, ideas, hobbies, aspirations, or addictions. Everybody wants to be part of one or more groups. These groups meet each other at home, in bars, or at other social occasions, and sometimes even at work. The search for equals and the communication that springs from it offers companies a number of opportunities to set up a portal that functions as a marketplace for those searching for information on a topic or searching for equals. This goes for business-to-consumer as well as business-to-business applications.

Vitual groups of people that meet via electronic media are numerous and cover a wide array of topics. As distance is not relevant, the reach of a virtual group is much larger than that of the local organization. Newsgroups, communities, and theme portals are virtual meeting places for people with similar interests. Children (and adults) no longer play computer games against the kids from school, but against kids on the other side of the world. Forming groups is a trend that occurs in many different ways among all possible target audiences. With mobile Internet, the virtual group comes closer to normal life. Communication among members is no longer limited to the evening hours in attics or during office breaks. The virtual group brings PC interaction closer to people that fear technical problems or do not have time.

Mobile Internet can add a dimension here. Wouldn't it be nice to show your vacation pictures to your friends on the screen of your mobile phone? Or the latest picture of the grandkids to your bridge friends? It is convenient to find out with mobile ICQ if you can call your friends on their mobile phones without disturbing them. Games will no longer be played at home, but also on the bus, at school, or at work. With mobile Internet, the virtual society can link with the real world. Mobile Internet will allow you to react faster to information from your direct environment. With mobile Internet, the high-tech image of many Internet trends will disappear, as it will be used in everyday situations.

Examples of adaptations that target group forming are the following:

- Mobile games for multiple players (see Figure 6.2)
- Portals with information on events
- Portals with information about hobbies or sights that warn you when you are close to an event that is relevant to you (for example, a wine lover visiting a city for business purposes is made aware of a winery by a fellow wine lover from that city)
- Services that enable you to find out where your friends and family members are, what they are doing, and if they can be interrupted (similar to ICQ)

Figure 6.2
BotFighters is the world's first location-based mobile game, made by It's Alive. In BotFighters, the players locate and shoot at each other with their cell phones out in the streets, where mobile positioning is used to determine whether the users are close enough to each other to be able to hit.

Fantasy Adventures

People search for adventure and distraction to bring excitement into their lives. They are not real adventures that include danger. There has to be no risk, and it has to be manageable and ending. This trend applies to numerous groups in society, both rich and poor, young and old. The success of organized vacations is an expression of this trend. Another expression is the search for unique objects such as art, tools from a particular culture, or the original objects, antiques, and clothing of famous people. Mobile telephony contributes to this in various ways. GSM phones will allow you to call from remote places. This gives the caller a sense of security during these "adventures." This will also enable the caller to stay in touch with family and friends. Mobile Internet can only increase this sense of security as information can be accessed locally when traveling, telephone numbers can be looked up, and dictionaries can even be used to translate menus or bills (see Figure 6.3).

As a result, mobile Internet callers can catch up with news, sports, and financial information while traveling abroad. This is a much-needed service. Think about all the people reading their American paper while traveling abroad. The Internet also plays a part in making reservations. KLM Click&Go and QXL offer an auction for airline tickets. It can be annoying to learn that somebody offered $10 more five minutes before the auction's close. With mobile Internet this is no longer inevitable. The mobile phone can receive a message as soon as a higher bid has been placed, allowing for increasing the previous bid. As an alternative for a real vacation there is virtual reality. Both kids and adults lose themselves in playing Playstation or Nintendo games. The new generation of computer games is connected to the Internet to play against others

Trends in Society 155

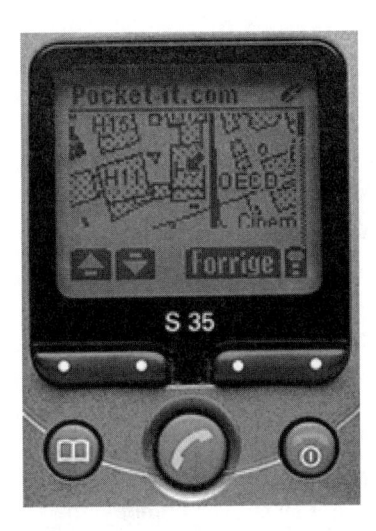

Figure 6.3
Example of a map service by Pocket It.

all over the world. Why limit this to the confines of one's home? Away from home there are plenty of opportunities to play a game. A third angle to mobile Internet is the impact it has on the search for objects. The Internet is a meeting place for suppliers and buyers. Electronic stores of new and used items, auctions, and search engines are all available. Mobile Internet adds a personal touch because you are the only one using a specific mobile phone. If you indicate what you are looking for, you can receive a message as soon as the item is found. Local relates to destination. As soon as your destination is available in the mobile network, it is possible to receive a message via your mobile phone. The item you have been seeking for years is for sale around the corner! The mobile phone gives you the address, telephone number, and directions to get to the store. Imagine this happening while you are in Bombay sightseeing! The one condition for this is that the customer is in control of that particular phone. Control is explained in greater detail in the next section.

The following are examples of applications that can accommodate fantasy adventures:

- Travel portals with local information on destinations
- Reservation sites that allow for mobile bidding on tickets and products
- Information portals with relevant information from home
- Search portals that help you find specific items

Egonomics

Egonomics relates to the individual. After a time of sterile computers, it is time for personal statements. It fulfills the need for personalization. Service is key for this trend, as reflected in customization, toll-free numbers, and 24-hour service 7 days a week. There is a demand for targeted services, especially in households where both partners work. Dogwalkers, clothing consultants, cleaners, and shopping services have been aimed their services at this group.

Personalized products such as pensions, mortgages, and computers or jeans made to order are also examples of this. The demand for customized products has resulted in a wide variety of models and shapes of products to meet these demands. An enormous increase in information and communication is the result. Mobile Internet will be an important instrument. In combination with the customization capabilities of a Web site, information telephone numbers offer many possibilities. Customers with questions about forms will want to contact the help desk right away. The combination of Internet and telephony gives customers the chance to express themselves.

The much-talked-about 24-hour society and its consequences for companies creates an opportunity for mobile Internet. Store after store extends business hours. More and more call centers can be reached on the weekends. The norm is changing and mobile Internet will accelerate this change. Mobile Internet is available 24 hours a day, 7 days a week and offers the necessary conditions for personalized services. Besides the fact that the mobile phone is an expression of someone's personality, the mobile phone offers access to a wealth of services. Besides calling, users can also surf the Net. This personal context offers possiblities for single transactions. The context makes cross-selling much easier, as it enables one to create an accurate profile of the user from a wide array of activities.

Whether a user decides to cooperate will be determined by what he or she receives in return. Online personalized news is offered by Yahoo!, CNN, and other suppliers. The user creates his or her own profile based on personal favorites. Headlines or interviews? The user determines this. News is only the beginning. More personal services follow: movies, recipes, games, dating, you name it.

With the growth of personalized services, privacy becomes a bigger issue. Users realize that after air miles loyalty schemes, bonus cards, and free Internet, their information is valuable. Handing over information is not a problem as long as the customer knows how it is used and what the benefit is. The benefit to the user may be financial, a higher service level, fun, or ease of use. The concern of privacy has resulted in people adapting their norms. If you just bought a new PC from a store, weekly flyers with special offers for PCs are not acceptable. Collecting air miles for two years has to yield something. Without direct positive results, the inclination to share information is short lived, response will dwindle, and the cost per new customer will increase.

Mobile Internet offers people the opportunity to tell their own story, not only via the telephone, but also via their own sites. On the Web sites on the Internet, all sorts of personal information is shared, complete with pictures and Web cams. Interactive personal ads or advertising; could it be more personal? Since April 2000, @info has offered users the ability to create their own WAP site, which can be accessed by everybody with a WAP phone.

Examples of applications that focus on egonomics can be found in every area in which available customer information can be used to improve service. Mobile Internet offers the tools to improve service and reduce cost.

The following provide a few examples:

- Services that prevent you from waiting an hour to be connected to the help desk after you went through numerous menus or make sure that you do not have to listen to a tape with the announcement that customer service is closed on the weekends.
- Insurance companies, banks, and other financial service providers, governmental agencies, the IRS, telecommunication companies, and Internet providers offer access to personal files for information inquiry.
- Services that help you search for a particular profile and warn you if such a profile is found, for example, realtors, employment agencies, used car dealers, or other supply-and-demand platforms.

99 Lives

By "99 lives," author Faith Popcorn refers to the trend of being busy, busy, busy. These days, everybody appears to be busy. Devices that make life easier are sold everywhere. The arrival of electronic organizers and mobile telephony is explained by this trend. The explosive growth of drive-through restaurants and TV dinners are a logical result.

This trend also caters to the mediators, who know personal preferences and give independent advice in complex purchasing situations. They also preselect products prior to purchase. A mediator is not cheap, but saves time. Examples are small companies hired by working individuals to select tasteful clothing or presents. Mediators in the financial service sector and independent mobile telephone stores fulfill this role. However, their independence is sometimes questionable, as a result of the premiums paid by suppliers. On the Internet, the electronic versions of these mediators are called smart agents. Search engines such as ASKJEEVES.com and portals fulfill this role. Several applications of electronic smart agents are in development. With mobile Internet the user can access his or her money at all times. Location and time of day will play important roles. Think of a reminder message to pick up dry cleaning before

6:00 p.m., or a message to go grocery shopping when you walk past the grocery store. As the smart agent is always in contact with the user via mobile telephone, the number of possible applications is boundless.

Another aspect of 99 lives is the increasing flow of information that people are exposed to via numerous media. The way that this information is dealt with has also changed. Channel-surfing behavior with commercials, and the fight against junk mail emphasizes the decreasing effectiveness of these media. Communication with specific groups in society is more difficult and therefore more expensive. The way people absorb information and the way they make decisions has changed. The classic communications methods are insufficient to communicate with the younger audience.

The ability to pay via the mobile phone fits the 99 lives trend. Time-consuming activities such as getting change for larger bills and waiting in line for vending machines and cash registers could be a thing of the past. Simplifying reservation processes will prevent waiting in line. This trend would also facilitate the combination of a phone and a calendar in a wristwatch. Work and personal time are much more closely related with respect to time and place. The mobile phone already plays an important part in organizing lives. The possibilities increase with mobile Internet: Think of reading email on the beach, checking if the server is still running at work from your couch at home, banking when and where you want to, ordering prescriptions, or ordering groceries from the store two blocks away from your office.

The introduction of the mobile phone as a fringe benefit gives the employer the option to enter into an employee's life outside of the office and after work hours. The employee is now able to build a workday after his or her own wishes. The employee feels obligated to carry a mobile phone around. The employer supplies the mobile phone with a particular goal in mind and expects that the telephone will be used. The use can be determined much more easily than it can for a PC. The employee is expected to be able to handle the device. A mobile phone could therefore to some extent play an educational role.

Examples of applications that facilitate 99 lives are the following:

- Services that help make reservations when you think of it, so that you do not forget it or have to wait on line at the grocery store, flower shop, clothing store, salon, dentist, or pharmacy.
- Personal agents that execute specific tasks for you. These agents are not fully automated: The human factor for the interpretation and selection of the right products is important. The agent will be available 7 days a week, 24 hours a day.

- Mobile access to different company-specific applications that should enable remote management and control without starting up your computer or hurrying off to a certain location. Think of managing automated company-critical systems and processes.
- Mobile payment systems for services and products, such as vending machines and parking meters.
- Logistic services provided to deliver goods to where they are expected instead of the user picking them up.
- Techniques that match customer profiles to relevant services.

The Vigilante Consumer

The consumer is more verbal than ever before with regard to products and services purchased. Writing a letter to a manufacturer, calling in a complaint, or informing a television show are no longer limited to a small number of whiners. Television programs give individuals the opportunity to be heard. Often, the masses are mobilized. Nonprofit organizations such as Greenpeace often play an important role. Many manufacturers have gotten into trouble as a result of public opinion: Think of manufacturers of tires, food, or insurance companies. The government is judged critically. There is a lack of confidence in companies and the government. False pretenses and a lack of leadership lead to mass protests and boycotts of products.

This trend affects mobile Internet in different ways. Both suppliers of mobile telephony and Internet service providers are often a topic of discussion among consumer groups. Both markets are not transparent, which makes a good comparison of quality and prices problematic. Many consumers complain about the services provided, unclear conditions, and the product itself. In mobile telephony, the battle for new antennas, radiation, and traffic safety are also discussed frequently. Many cities and people do not want mobile antennas on their roofs. Substantial amounts, often more than $5,000 are paid for a hot spot. Besides pollution of the horizon, the danger of radiation has been a concern about the antennas. Research costing $25 million has been conducted on the effects of radiation of mobile phones and antennas, but governments, users, and suppliers do not agree on the effects of mobile calling on the human body. British researchers have recommended limited use by children. However, they have not found new evidence indicating that GSM telephones are bad for your health. Dutch research institute TNO was also unable to find negative effects on the brain. The FDA has announced it will research the health risks, which will take three to five years and cost $1 million. However, the urge to have a mobile phone is so strong that many consumers disregard these risks.

The apprehensive consumer gets a new device as a result of mobile Internet. The Internet already contributes to the power of the consumer by giving the individual easy access to an unending number of options for discussion, judgment, and group forming. The influence the Internet already has is a sample of what is yet to come. As soon as Internet via the mobile phone is attainable for a larger part of the population, the options for the Internet will also be developed further. Mobile Internet gives critics a new, massive, and very direct medium.

Another result of this trend concerns the ways of interaction on the Internet. Now that the Internet has become a mass media, the need for rules and regulations arises. The united need for Net etiquette is obvious. Spam email, email bombs, and sex content filters are all expressions of these needs. Mobile Internet suppliers will have to give users the confidence that their privacy is secure, that their personal information is treated confidentially, and that their personal lives are not violated. The strengths of mobile Internet are the personal and interactive character of the medium combined with the place- and time-independent availability. This strength becomes a weakness if treated carelessly.

Examples of applications that benefit the apprehensive consumer include the following:

- Services that help customers make purchasing decisions with information about manufacturers, ingredients, prices, and alternatives
- Portals that mobilize dissatisfied or weary customers
- Portals that poll customer opinions on specific topics (a 10-minute representative market survey)
- Services that enable security, filters, anonymity, and other privacy-oriented activities

Female Think

Female think is teamwork instead of hierarchy. It is asking the right questions instead of the need of finding the answers. It also means more parts and opinions versus a single part or image. Female think also includes adapting to change instead of resisting change, process thinking instead of goal-oriented thinking and relationship-oriented thinking instead of transaction-focused thinking. The trend of female intuition is interesting not just because the female intuition creates many opportunities for new services, but because this trend provides insights that form important conditions for the success of a technological development such as mobile Internet. This trend is not only relevant to reaching the female audience; it also appears in environments dominated by men. Think of the growing interest in team spirit and teamwork, and the value attached to people who develop and maintain strong relationships other than on a hierarchical basis. The continuing change in large corporations and the market entry of some small dynamic companies are also examples of this trend.

Mobile Internet is a medium that does not comply with hierarchy. Mobile Internet, as mentioned before, will bring about big changes. From this perspective, mobile Internet links nicely to this trend. On the other hand, this trend could undermine the success of mobile Internet. The danger lies in the way it is used. Currently, WAP is mainly the domain of technologists. The threat is that WAP may become a goal instead of a means of interaction. Services and mobile phones are then no longer developed from the perspective of optimizing the interaction process. Technology-oriented services will be completely unattainable for laymen. PCs are still dominated by technology. As a result, the Internet is still limited to a selected part of society. Mobile phones have the potential to break through the technological dominance of the PC. Internet for the mass market seems near in this scenario.

Crucial to the success of mobile Internet is the extent to which services are aimed at building relationships. Mobile Internet offers the possibility of personal access to a user. This is confirmed by the users of mobile phones. Look at the careful way the daily paper generation gives out their mobile phone numbers. Also, the increasing number of complaints about unwanted SMS messages is an indication that mobile telephony is a much more personal medium than the Internet. Companies that use mobile Internet to generate transactions will appear aggressive to users. This decreases the chance for success in the long term. There is no investment made in the development of a relationship. Companies that invade users' privacy (from their perspective) often do not get a second chance.

The trend of female thinking creates conditions instead of opportunities. Important conditions include the following:

- Discussion and opinion facilities
- Facilities that make the relationships between people and between people and companies stronger
- Facilities that integrate the various means of communications: telephone help desk, (mobile) Internet, physical branches
- Attention to ergonomics and simplicity, making the user the center of attention with the design instead of the product

Cocooning

Cocooning is a trend in which people pull back to the confines of their own home in their own neighborhood. Cocooning is enforced by fear that it is not safe to be outside. A sense of security keeps people inside their homes at night and sometimes even during the day. Cocooning results in high expenditures on home and garden. This is an important trend, especially for families and the elderly. Mobile telephony and cocooning appear to be opposed, but this is not the case. Buying a mobile phone gives many people a sense of security when they are on the road. People do not like to be surprised by cars breaking down or a lunatic on the street. A mobile phone provides a way to reduce this insecurity and check up on things. Down the road, this trend will develop from shopping via the computer into shopping via the mobile phone, as long as the customer gets a receipt for paying. You get exactly what you want, without having to leave the house. You order only from the companies you trust, of course. The mobile phone can offer this group more security than the Internet.

If you have to leave the house, it is easier to keep everything under control. Think of a Web cam in the nursery. You can keep tabs on the baby via the mobile phone. You will also be able to control appliances in your home remotely: turn on the lights, check to see if the alarm is turned on or if the stove is off, turn up the thermostat before you get home, or switch on the oven when you are on your way home. You will be able to consult recipes, find tricks for removing stains, learn about treatment for plants or animals, or find afterschool help for your children via television or (mobile) Internet, but also via appliances such as refrigerators (Figure 6.4) and washers with Internet access.

Figure 6.4
Electrolux developed a refrigerator with an Internet screen to search for recipes online, among other things. (See *www.electrolux.se/screenfridge/*)

The trend of cocooning offers opportunities for services such as the following:

- Locating, ordering, and paying for services and products via the Internet
- Mobile control of household appliances such as ovens, heating, stoves, lamps, washers, locks, and alarm installations when the user is away from home
- Applications for which mobile phones can be used as a remote control inside the house, making use of Bluetooth where possible
- Services that allow the user to inspect products from a distance
- Portals with detailed information on home, garden, family, and hobbies for which mobility provides added value (for example, a Q & A session that can be accessed when you are in the middle of home improvement chores)

Small Indulgences

This trend stems from the feeling that security subsides. Securities such as pensions, Social Security, and lifetime employment are constantly subject to cuts. The feeling is that you try hard, but you cannot change the world alone. Small favors are a present or reward for you or somebody else. They depend on the person. Gourmet food, chocolate, or flowers are good examples. In high-income groups, small, smart, exclusive, or status-increasing high-tech gadgets are favored, such as a high-tech vacuum cleaner, an electronic organizer, or a new mobile phone. To buy something nice for yourself and enjoy it are things people are less ashamed of these days. It does not have to be expensive, as long as you enjoy it. If enjoyment can be bought at a reasonable price, even better: Buy smart and buy cheap. Mobile Internet offers sufficient opportunities for manufacturers of special models of mobile phones. The success of the small silver Nokia 8850 and the Motorola V-series have proven this. There are plenty of opportunities for suppliers of services that help the user locate products for a reasonable price. Letsbuyit.com, priceline.com, and numerous others are examples of services on the Internet that are part of this trend. Services that indicate the closest location to purchase a product or service make the satisfaction more easily attainable. With mobile Internet, you can give into the need to satisfy personal needs. You can call for these services whenever and wherever you want.

Examples of applications that facilitate the small indulgences trend include the following:

- Services that offer impulse buys at the right time and the right way, including delivery or directions
- Services that offer product comparisons, auctions, or discounts

Cashing Out

By cashing out, we mean choosing quality of life over a top career with a multinational corporation. Experienced managers resign and start their own companies or work from home. Students found their own Internet startup company right after college. People try to avoid traffic jams by working from home. Cashing out also includes the increase in sales of SUVs, casual Fridays, and parental leave.

The disappearing structure and hierarchy of large companies creates a need for networks. Smaller companies find each other as they complement each other. They have a common interest in spreading knowledge. Portals with information and services in the area of technical support are being developed; for example, Web and WAP site construction, maintenance, and hosting; legal services; financial services; market information; or business consultancy, to name a few. The Internet is a suitable medium for these services. Due to the limited size and external focus of these companies, mobile access to this information will become necessary (see Figure 6.5). Coworkers are often not in the office but still need access to information.

Figure 6.5
Access to important information, wherever you are.

The reduced barrier to market entry and the numerous possibilities for partnership make it relatively easy for employees to start their own businesses. The number of Internet startups in Europe will increase as a result of mobile Internet. In Finland, Nokia has surrounded itself with a large number of small, specialized, high-tech companies. For larger companies, working outside of the office and after office hours has consequences. It also requires communications facilities to access the company network from an outside PC or laptop. Mobile access to information on the company network is, of course, necessary.

Examples of applications that facilitate the cashing out trend are as follows:

- Services that help small companies with their business in financial, legal, or technical areas
- Mobile access to files that were only available within the physical company, such as price lists, customer databases, Intranet and telephone directories, online calendars, and email
- Communication and teamwork services for groups of allied companies

Icon Toppling

This last trend relates to the dwindling belief in systems and institutions. Faith in systems and institutions determined our thinking for years. Belief and faith in government, companies, police, and marriage are under pressure because expectations have not been met. Corrupt politicians, never-ending procedures, powerless police officers, hackers, and reorganizations that impact employment are a few examples that clarify this trend.

Where expectations are not met, opportunities arise: opportunities to make life easier for customers and employees, opportunities to increase customer satisfaction levels, and opportunities for large companies to regain trust by being present in the local market and using local events to build relationships. A very good Dutch example is the sponsorship of the new year's dive in the ice cold North Sea by Unilever sausage brand Unox. Mobile Internet offers opportunities for small companies to do better than larger companies. The Internet has shaken industries to the core, and it still does. Personal character and local presence are means that small local entrepreneurs have at their disposal and mobile Internet offers these opportunities.

START NOW OR WAIT

The consumer trends just explained were given by Faith Popcorn, a renowned American futurologist. In her practice, she tests every idea or development against 16 criteria of trends. She uses three criteria to judge if the idea can be successful:

- It has to satisfy a long-term need, and not be shortlived hype.
- The target audience has to be large enough to determine longevity.
- A development has to be applicable to at least four trends to be successful.

Mobile Internet meets all three criteria. As mentioned in Chapter 1, all important parties in the area of information and communication technology are supporting mobile Internet. This indicates that mobile Internet is not shortlived hype, but a next step in the converging worlds of Internet, media, and telecoms. Chapters 3, 4, and 5 explained developments in penetration of the Internet. Mobile telephony and mobile Internet indicate that the market is large enough for a company to develop new applications. Mobile Internet links to 11 trends, especially cocooning, clanning, small indulgences, egonomics, and 99 lives.

With regard to the trends, a translation of services provided for the Internet to mobile Internet is obvious. The services for mobile Internet will be more elaborate, as the unique character of mobile Internet allows. Internet experience is gained from e-commerce sites, auctions, and sites for bargain hunters or e-zines. Knowledge is gained by bringing together consumers with similar interests, or by facilitating communication between people with similar interests. This is also applicable to mobile Internet. Services will become available online and more current via the mobile phone. Integration with speech communication will lead to new applications. The investment in permission marketing, building customer profiles, and personalized offers will yield more with a personal medium such as mobile Internet. It remains to be seen if the "everything is free" business models also work for mobile Internet. Users are probably willing to pay for current personalized news on demand via their mobile telephone. Mobile Internet will decrease identification and payment problems, so that it will be possible to charge users fees.

If the location of the user is also known, new opportunities arise, which we can only imagine. The hard-to-ignore fast food offer on your mobile phone when you walk past a McDonald's seems strange and unwanted now, but our perception is likely to change over time. If the company makes the offer substantial and the user can select whether or not to receive the offer, it all sounds less intrusive. City maps with search services are so trivial that it is very likely applications with surprising business models will appear.

Location determination will function as a bridge between the virtual world and the real world. With mobile Internet and location determination, Internet services will not only be available in the study room or in the office, but also in the streets. This has far-reaching consequences for store owners. Price transparency increases, resulting in customers wanting to know prices at different locations. Customers will decide if they want to walk two more blocks for an extra discount. This has a certain effect on prices. Competing stores are probably even closer, less crowded, and have free parking. This not only affects the big boys in retail, but also the local store owner. Differentiation on aspects other than price is linked to increased efficiency.

The opportunities the Internet offers for niche marketing also apply to mobile Internet. Niche marketing links to the need for more service. Locally, a niche may not be worthwhile, but worldwide niche markets offer great potential. User profiles offer enough to really reach those interested. The strength of many local entrepreneurs is personalized contact and customer knowledge. The threat of the Internet is relatively small for most local entrepreneurs, but this could change for mobile Internet for the masses. Fortunately, the personal interaction with the store owner differentiates him or her from an Internet site. The question is how much of a price difference the consumer is willing to pay for this.

Loyalty will become more important. Mobile Internet becomes a valuable instrument for loyalty programs. As the mobile phone can be used in many different ways, the possibilities seem endless. If user information can be used, an accurate user profile can be built. This becomes an important weapon in the battle against price erosion.

Brand building is even more important for mobile Internet than for regular Internet. Mobile is an instant medium with which information can be found rapidly. Navigating through menus is more difficult than it is on the Internet. Users will bookmark preferred services and search for familiar brands. These are brands from the real world, the Internet world, or perhaps the Amazon.com's of the mobile Internet world.

The developments surrounding the Internet have taught us that speed is important. Existing companies have to invest heavily to go through the learning curve. First, movers will also try to take the lead over established companies with mobile Internet. Even if a company has plenty of expertise in the area of the Internet, there still remains plenty to learn to become successful with mobile Internet.

Businesses are interested in developing mobile Internet applications. Regardless of whether the services companies offer are focused on consumers, companies, or their own employees, there is much interest from different branches of industry. American success stories have been achieved already using Palm's Web clipping. A few examples: Famous Footwear eliminates 75 percent of pricing errors by giving stores handhelds with scanners and wireless links to the corporate network. AMR, a large ambulance company, captures patient medical data prior to treatment, cutting costs and saving lives. Mitel Corporation, a telecom company, lets its executives access and update vital business information quickly and effectively. Volvo reduced turnaround time of quality inspections from three weeks to one day, realizing huge cost savings.

Starting with mobile Internet does not have to be linked with large investments. In Chapter 2, we explained that the differences between WAP and the Internet are limited when it comes to technology. Hardware and software purchased for the Internet and the technological knowledge of personnel can be used to lower startup costs. Investments in back-office processes and logistics that benefit the Internet can also be used for WAP. The costs should not keep you from WAP. Maybe your employees are already experimenting with this technology.

MOBILE INTERNET IN BUSINESS-TO-BUSINESS RELATIONSHIPS

The description of trends is mainly applicable to the consumer market. Market research companies such as Gartner and Forester have high expectations of business-to-business developments. Even though the examples are not as outstanding, we do not want to pass them by. B-to-C services brings about new B-to-B relationships. All consumer products have to be manufactured, collected, packaged, delivered, and invoiced. Web site construction, telecommunications, marketing and communication, and after-sales service generate turnover for suppliers and partners. Many of the consumer applications can also be used in a B-to-B setting. Think of traffic information, reservations or payment for parking spaces, hotel rooms, restaurants, airline tickets, or pizza.

It does not stop. In a number of mature industries, marketplaces for e-businesses have already been set up. Market research company Gartner has calculated that these marketplaces will generate 37 percent of B-to-B commerce in 2004. This marketplace targets a specific sector or territory. The car industry takes the lead with a platform for suppliers and car manufacturers. The cooperation between Oracle, General Motors, Daimler Chrysler, and Ford offers so much promise that others within the industry want to join. This was enough cause for the FTC to start an investigation into the monopolistic position of this cooperation. Virtual marketplaces also appear in the energy, plastic, metal, and iron industries. The objective is to gain an advantage in cost by matching supply and demand more closely and by communicating worldwide in an easy and cheap manner. Why should these initiatives be limited to the regular Internet? A recent development in this area is the founding of a marketplace for construction materials, which can be accessed via the mobile phone. Just-in-time orders for materials are essential in continuing operations. What could be easier than ordering materials from the construction site via the mobile phone? For many applications, an in-depth analysis is not needed. Ordering products via the mobile phone prevents hassles and delays. This applies to commodities such as office supplies, hardware and software, office furniture, and so on.

Besides the relation toward customers, the internal information supply within companies and between cooperating companies changes. Internal no longer means physically inside a building. Inter-, intra-, and extranets, email, electronic newsletters, knowledge databases, and group directories have completely changed the daily activities in the office, and this will continue. For the employee at home, telecommuting has come to fruition. The employee on the road was often not involved until now. The development of company information systems for mobile workers is an important trend. WAP telephones, laptops, Palm Pilots, and PDAs will be widely used for business. Sales representatives can check on prices and deliveries while they are on a sales call, consultants can access information while with a customer, and managers can inspect sales numbers during the weekend. These are just a few applications that can lead to efficiency, higher profitability, and higher customer satisfaction.

MOBILE INTERNET AS TREND ACCELERATOR

Mobile telephony, Internet, and soon mobile Internet will play a part in the creation and distribution of trends. A trend is basically nothing but news. A newspaper brings news seven days a week, but it is at least half a day old. The television brings the latest news several times a day at set times. The Internet always brings up-to-date news as soon as you log on to a PC. Mobile Internet gives you direct up-to-date news, where and whenever you want.

Internet and mobile telephony have increased the ease and the speed of communication, just like television did years ago. Email and Internet make fast, direct interaction with a large number of people easier. Interaction and communication are both necessary for the creation and distribution of trends. Speed does not directly relate to the number of bits per second of the connection. We refer to the speed with which a message can be sent and the speed at which information can be accessed. Sending flyers by mail or by fax cannot compete with this. Also, the distribution of trends via social interaction from city to city and from country to country could take months or even years. By direct messaging we mean that the recipient of the information receives the message himself or herself. Unlike direct mail or television commercials, the information is displayed on the screen when the user wants to read it and how he or she wants to read it. Communication with a large number of people is not so special by itself. What makes the Internet and email special is that a large number of people can be reached at a lower cost.

Distribution of trends goes beyond sharing information. Trends often go together with complementing products or services. Think, for example, of a trendy outfit. Trends that spread via the Internet form an ideal context for the sales of the products and services related to that trend. With the sales of the products and services, the trend becomes more visible in the real world. This visibility will intensify the trend. People sensitive to trends will notice the trend in the street and search for products via the Internet. Of course, accessories can be bought right away.

Mobile telephony allows people to be reached if they are not home or are away from their desk. This makes direct communication between people simple and more intense. Trends often spread via groups of people with similar interests, people who know each other, or people who have something in common. Intense communication made possible by email and mobile telephony accelerate the distribution of trends. Mobile Internet makes the Internet accessible in places and situations where you will not sit behind a PC. It brings the Internet to the people. Mobile Internet will give the distribution of trends a new impulse for three reasons.

First, the trend is not only available via the Internet in the study, but also via meeting places where trends are made and distributed: school playgrounds, clubs, or cafés. During a conversation you use your mobile phone to illustrate what you mean. You search for information, or you mail a URL to your friend's phone. Using the mobile phone this way offers a spinoff in the shape of word-of-mouth advertising. People tell each other that there is great information on a site, and they can show it right away. This happens worldwide and has a positive effect on the use of the mobile phone.

The second reason why mobile Internet will spread at an increasing speed relates to the sort of information to which trends belong. Unimportant information is often viewed as littering. People often read a flyer or magazine while waiting in a store or at the dentist. Free newspapers owe their success largely to the need for reading materials during a train journey. There are many more of these moments we can think of. Access to this kind of information via mobile phone seems to be right on the money.

Finally, information on trends comes within reach of people that hardly use the Internet at all; people who are trendsetters or trend followers in different areas. Trends can therefore be noticed and followed faster by large groups of people. As other people also use the medium, the number and diversity of trends via the Internet will also increase. In that case, mobile Internet will not only expedite the main technological trends, but also the trends in, for example, sociological and cultural areas, such as fashion or gardening. Mobile Internet also offers opportunities for local initiatives because there are plenty of people that have access to the Internet via mobile phones within smaller local communities.

7 Opportunities and Threats of Mobile Internet for Your Business

In this chapter...

- Customers   173
- Competition   177
- New Entrants to Market   179
- Substitutes   183
- Suppliers   188

As discussed in the previous chapters, mobile Internet offers numerous opportunities. Crucial to the success of the mobile Internet is the preparation of the service provider. These preparations are in full swing for some innovative companies. One after another, public companies announce large investments in new media and acquire small companies with specific knowledge. All renowned consultancy companies are delving into e-business and m-business. E-commerce platforms by HP, IBM, and Sun will provide access to WAP, even if this is not yet possible. Key are the different ways of communication with customers, which will enable customers to make their demands known. It is all about systematically building the knowledge of customer preferences and habits. This knowledge is used in communication with customers to better anticipate needs and in the acquisition of new customers. It is about a different way of production and distribution of products and services for customers as a result of specific needs, different business hours, new distribution channels, and the need for higher levels of efficiency. It is about new and unexpected competition or existing competition with a new competitive advantage. It may even be about a whole new business model.

In Chapter 6, we indicated by example which services could be offered with mobile Internet, keeping certain trends in mind. In Chapter 5, we looked at the conflicts that could arise in the market for mobile Internet. This chapter focuses on how to estimate the opportunities and threats of mobile Internet for your company. The opportunities of the mobile Internet are analyzed using Porter's strategic competitive analysis. To complete the analysis, specifics of the regular Internet are taken into account. As there is plenty of literature covering the area of the regular Internet, we do not claim to cover the topic of mobile Internet completely. Examples where mobile Internet offers new possibilities stem from the regular Internet, expanded with unique characteristics of the mobile Internet. These are availability, methods of payment, directness, position, and personalization options.

The entrepreneur wanting to use the mobile Internet to conduct successful business is the point of focus here. This entrepreneur can be active in any industry. The five pillars mentioned in Porter's competition model are the following:

- Position of the customer
- Competition
- New entrants to the market
- Substitutes
- Suppliers

In Chapter 6, we covered the trends and the relationship between supplier and buyer. Research conducted by *Information Week* among managers in the United States shows that the most important motives for e-business are related to customers and competition (see Figure 7.1). The threat of new entrants and suppliers to the market

What Drives E-Business?
Which business goals were highly significant in your organization's decision to deploy electronic-business applications?

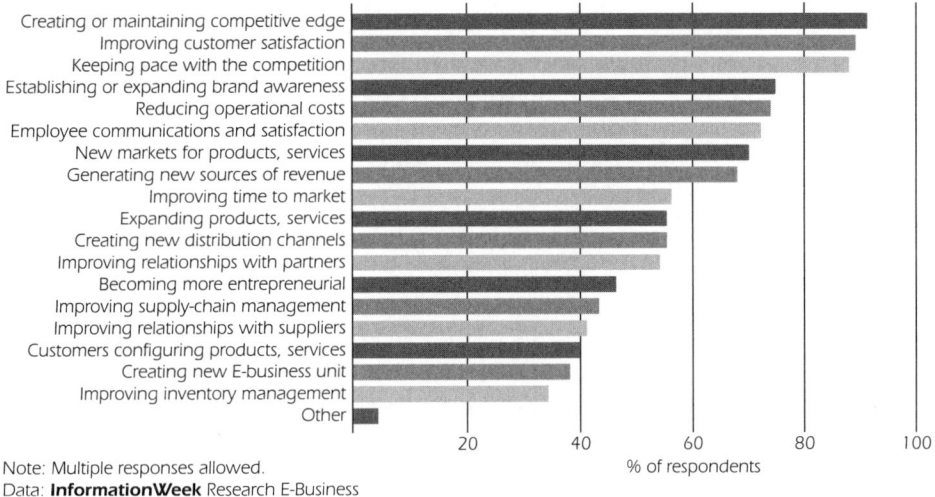

Goal	% of respondents
Creating or maintaining competitive edge	~90
Improving customer satisfaction	~88
Keeping pace with the competition	~85
Establishing or expanding brand awareness	~75
Reducing operational costs	~73
Employee communications and satisfaction	~72
New markets for products, services	~68
Generating new sources of revenue	~67
Improving time to market	~58
Expanding products, services	~55
Creating new distribution channels	~55
Improving relationships with partners	~52
Becoming more entrepreneurial	~48
Improving supply-chain management	~45
Improving relationships with suppliers	~42
Customers configuring products, services	~42
Creating new E-business unit	~40
Improving inventory management	~37
Other	~5

Note: Multiple responses allowed.
Data: **InformationWeek** Research E-Business Expo Study of 375 IT and Business Executives

Figure 7.1
Main reasons to start using e-business applications.

cannot be ignored in the battle for the best place in the mobile value chain. Substitution of existing distribution and communication channels by mobile Internet also has to be taken into consideration.

CUSTOMERS

Transparent Information

Mobile Internet will increase consumer power as the supply of products becomes more transparent. Supply of information is linked to Internet mechanisms as price comparison and buyer concentration. The improved supply of information clarifies the terms to the customer. Small print in contracts becomes more visible, clarifying the switching costs, terms of contract, and guarantees prior to purchase. Clear terms give suppliers an advantage in the selection process of the customer. Price comparison via the mobile telephone will start a revolution.

Jaysar offers the ability to compare prices for books, CDs, DVDs, and games via WAP. Buyers concentration, like with United Customers and Let's Buy It on the Internet, increase buying volume. These initiatives can also be developed for mobile Internet. The improved supply of information allows for a clearer view of the differences between products. Functional differences can be compared right away with price differences, putting prices under pressure. The influence of mobile Internet will not be this strong right away. There will always be people who do not want to put any effort into comparing products. Finally, there are sufficient suppliers with services that are difficult to compare via a mobile phone and for which personal contact with salespeople is the determining factor.

Mobile Internet will be used by customers on the street and in the store. For this reason, the impact of mobile Internet for "brick-and-mortar" retailers is potentially much larger than with the regular Internet. The current problem of Internet stores is that they cannot deliver the goods right away. Because mobile Internet allows the customer to be in the physical and the virtual store at the same time, he or she can search the mobile Internet for a product or a store, or determine location. Next, the customer can buy the product right away and take it home.

Improved Customer Service

For stores, the mobile Internet offers an opportunity to improve the level of service for customers if they give customers access to the inventory database, allow them to make reservations, and have goods ready for pickup. If payment is settled, too, the customer only has to stop into the store to pick up the product. Also think about takeout meals, which you can order via the mobile phone on the train and pick up at the station, without having to wait another 30 minutes. The possibilities UMTS will offer with, for example, mobile video (Figure 7.2), go far beyond that. Parents will then have the opportunity to view their kids at day care via the mobile phone.

Figure 7.2
A video camera on a mobile phone is possible with UMTS.

In the street, customers can—if they want to—be made aware of the special offers of the stores they are walking past. The brick-and-mortar companies can use the mobile Internet to lure the customer into the store, and once inside, help the customer find the right department. This will make shopping easier and more enjoyable. If the customer enjoys this way of shopping, the barrier to switching suppliers will be much higher.

A mobile telephone can also function as the carrier of a loyalty program, as a method of saving, or in the form of targeted information. There is a WAP site that hosts 160 hair salons, including information on hair-care products. The possibilities for targeted services can contribute to a strong customer relationship. Travel agencies can also provide services during a trip. The customer is able to check up on all his or her personal travel information, and will receive messages if there are delays or changes in the itinerary. The customer can ask for assistance right away if need be. Shows can differentiate themselves by making information available to visitors before, during, and after the show. In sports stadiums and at concerts, big screens can be used for communication to the fans, for example, raffles, chats, and bulletin boards. The people at home can view this via the Internet or mobile phone.

Payment

The development of methods of payment is a problem area for many financial service providers and telecom companies. From the customer's perspective, a payment facility is more a necessity than a differentiating benefit. Payments and identification via the mobile phone contribute to the reduction of cost, for both small and larger payments. The Dutch company Schmidt offers customers the ability to reserve and pay for a parking spot via mobile phone. In industries where direct payments are less common and larger amounts are involved, or where the chance of fraud or robberies is large, the potential cost savings are even larger. Think of ordering pizza and paying for it while you order over the phone; there is no need to have money on hand and the delivery staff runs less risk of getting robbed. Also, limiting terms of payment by using mobile payments and simple identification are examples of possible cost savings. A warning message of the due date on bills can reduce collection agency costs considerably if the option to pay via the mobile phone exists. However, the acceptance of customers is important, as well as legal issues. If a paper invoice still has to be mailed, the cost savings will be limited.

Viewing invoices is often an important need for customers. Checking the status of payments via the mobile phone or the Internet can satisfy any such needs. Obtaining clarification for certain transactions or possibilities for internal cost allocation contribute to customer satisfaction and reduce the number of phone calls to customer service.

Identification may become very easy with the mobile phone. An employee normally has to identify himself or herself and let a boss sign to be able to buy on the company account. With the mobile phone, identification and approval of payment is much easier, while the regular invoicing process remains.

Impulsive Actions

For suppliers of services with a limited capacity, such as airlines, cinemas, theaters, and car rental companies, mobile Internet offers great opportunities. Reservations can be made and paid in advance, decreasing the number of no-shows. Those interested can sign up on a waiting list and be called if tickets become available. This ticket can then be purchased at a lower price (see Figure 7.3). This enables people to decide if they want to go see a show that evening without having to stand in line for tickets. The same principle applies to art exhibits, the circus, concerts, games, or restaurants.

Table 7.1 summarizes opportunities and threats of mobile Internet from the customer's perspective.

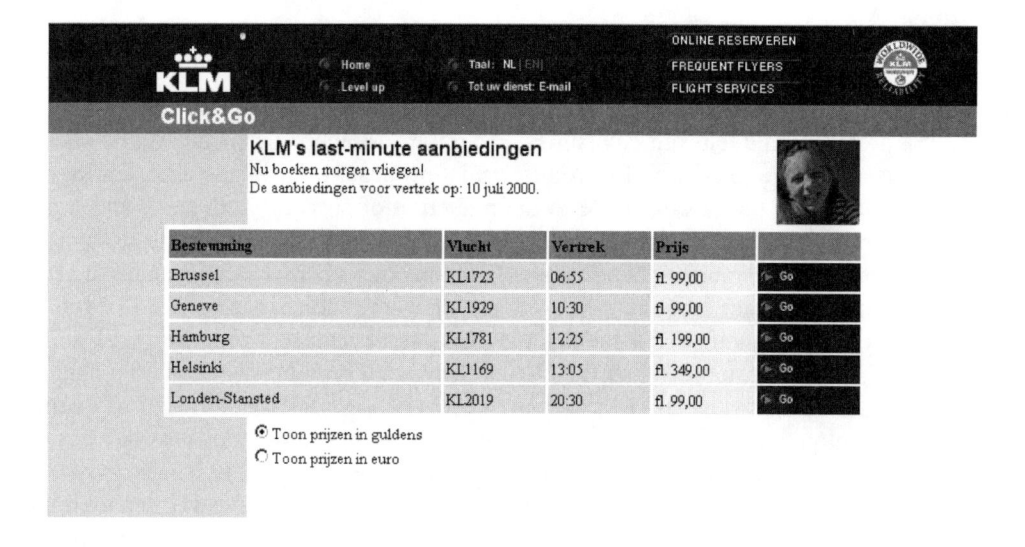

Figure 7.3
With KLM's Click&Go you can book airline tickets at the last minute.

Table 7.1
Opportunities and Threats from the Customer's Perspective

Increased transparency and product information provides customers with insights into product differences, conditions, availability, and price levels.
Buyer concentration and real-time price comparisons are available for the masses and in the regular world. The customer can choose on the shop floor whether to buy the product right then and there, in the store around the corner, or via the Internet with home delivery. This puts pressure on prices for products and services that can be compared easily.
Servicing new and existing customers by providing better information, ease of payment, or clear conditions.
Improved customer relationships by personal services wherever and whenever the customer wants.
Reduction in costs and increase in customer satisfaction by offering electronic payments and electronic invoicing.
Higher levels of capacity and extra services for suppliers of services with limited capacity.

COMPETITION

Mobile Internet can be used in different ways to stay ahead of the competition. In the previous section, we mentioned using the mobile Internet to improve the service levels for customers. Improving service levels can be done in different ways.

Ease of Use

Many people do not want to call for appointments with companies. People will have the intention for weeks but they constantly forget to call or give up when they are put on hold. Many companies that offer services for which the customer has to make an appointment can make it much easier for their customers (and those of the competition). Think of annual car maintenance appointments. Setting up appointments for regular events such as dentist visits, hair salon, or beauty salon can be scheduled via the mobile Internet if the user instantly wants to. The dentist can also send a mobile reminder instead of a card. Appointments with financial service providers such as tax consultants, mortgage providers, insurance companies, and tax experts can also be made, allowing for more effective use of time and providing customers with advice faster.

"In-Store" Competition

In the streets, people can be persuaded with offers via SMS messages if they are open to that. Mobile Internet also offers the chance to compare products in the store right away with products in the corner store or a virtual store. This allows WAP suppliers to be in the competitor's store. It will change the shopping experience if customers use their mobile phones to get a second opinion after they speak to a salesperson. Window shopping and buying at a different location is a possible result. Many products are not ordered electronically. In the orientation process, customers will still visit stores, but the actual purchase can take place electronically. The current trend to turn shopping into an experience makes it necessary to compete in areas other than price. Mobile Internet can be an instrument to enhance this experience or to develop the experience outside of the store. In the large sporting goods stores, for example, customers and other interested people can receive a message when the basketball demonstration will begin.

"The place to be" can be distinguished via the mobile Internet. Event listings, concerts, menus, and business hours can all be offered via WAP.

Additional Communication Channels

Mobile Internet can be used as an additional communication media, in addition to local newspapers, advertising, and Internet sites. The impact of mobile Internet varies in our opinion. The regular butcher's customers will not go to a different butcher all of a sudden. For many quality goods such as fresh products and clothing, the impact will be limited. For competition-sensitive products and fads this may be different. Beer brands can, in contradiction to the virtual bar on the Internet, take their advertising campaigns one step further by hosting real bars in their WAP services, with real beer and real people. Price determinants have already been discussed: The claim "lowest price guaranteed" has to be proven.

Table 7.2 summarizes opportunities and threats of mobile Internet from the perspective of competition.

Table 7.2
Opportunities and Threats from the Perspective of the Competition

By eliminating barriers such as reservations, making appointments, and boring telephone conversations, ease of use is increased for the customer, resulting in higher levels of customer satisfaction. The service provider can differentiate itself in a fairly simple manner.
The direct availability of mobile Internet offers the opportunity to pull customers from the competitor's store.
Using the mobile Internet as an instrument of communication in price and trend-sensitive branches could lead to increased competition and decreased loyalty toward the service provider.

NEW ENTRANTS TO MARKET..................................

Borders Are Less Relevant

Just as with the Internet, borders are less relevant. Portals aimed at a specific target audience can operate at an international level in which global niche markets form a monster market. More and more software is available for the Internet to differentiate sites by culture. Differences in languages, look and feel of a Web site, and even product line can be realized at a relatively low cost. This leads to efficient targeting of local markets. New competitors can enter the market easily. Reputable names from other countries make their way across borders faster. Even the local entrepreneur has to face a variety of competitors, some with completely different business models than were the norm up to now. Mobile Internet comes much closer to the physical store, as the customer has the mobile phone on hand while shopping. As a result, expansion to other countries becomes much more attractive.

Banks can offer their Internet and WAP services in other countries without the obligation to set up local offices. Virtual banks and investment companies already exist. Many people have an investment account with E*TRADE by Schwab. Schwab realized early on that the Internet offered great opportunities for further expansion. Schwab anticipated this and now services worldwide customers with investment accounts via the Internet (Figure 7.4). Mobile Internet gives you the option to use bank

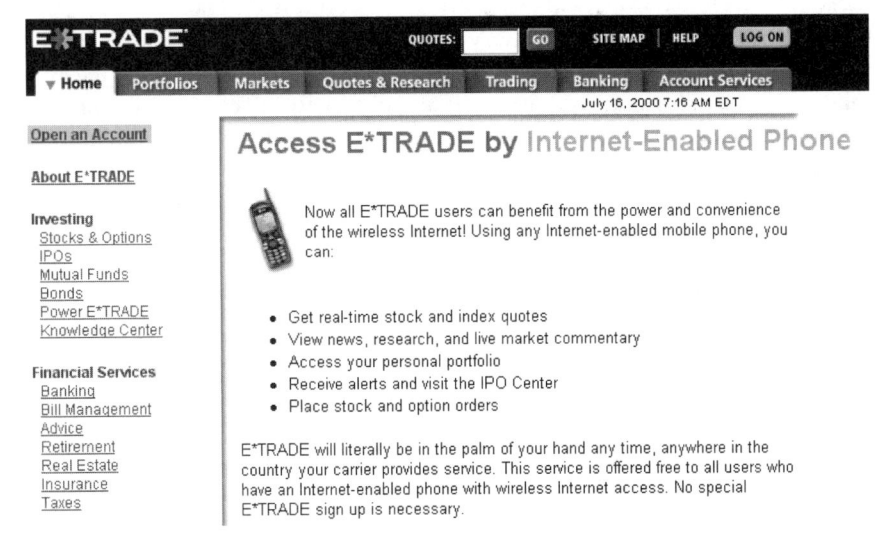

Figure 7.4
E*TRADE offers stock trading via the Internet and the mobile Internet.

services anywhere, not only at home behind the PC. This gives investing an added dimension. Other bank services such as paying bills or underwriting insurance can be done right away without filling out forms or waiting in line. A person interested in buying a house can view online how much the monthly payments will be before placing a bid.

The absence of a physical office network allows for cost advantages. These can be passed onto the customer, for example, by offering lower rates for purchasing stocks. The majority of the American, European, and Asian bankers find cost efficiency and consolidation more important. According to research conducted by Arthur Andersen, bankers see e-business as a channel for information, comparable to letters and leaflets. The fact that two-way communication via the Internet and mobile telephony provide a solution to their problems does not seem to have registered everywhere. There are opportunities for smart entrepreneurs in the financial sectors!

The Pioneer's Advantage

By entering the market with a low threshold service, a company can develop a brand in a new market in a short period of time. American Internet pioneers have proved this. Amazon.com has brought the ease of ordering books via the Internet to Europe and has taken a lead on its European competition. Amazon.com has not had the need for television advertising to draw customers, unlike BOL. Amazon.com allows other Web sites to benefit, as soon as they link their customers to Amazon.com to buy products. It looks like Europe will now have the opportunity to take the lead over the United States. As a result of the high level of mobile penetration in Europe, the availability of one standard, and the enormous push for mobile Internet by the telecommunications industry, the European mass market appears to be ready for the mobile Internet sooner than the American market. The large American Internet companies have noticed this too. U.S. companies such as Yahoo!, Alta Vista, and Amazon.com have founded new business divisions focusing on Europe and on mobile Internet.

Brand Power

Whether the actual number of suppliers will increase is doubtful. The concentration processes in many industries and the trend among multinationals to cut their brand portfolios reduces the actual number of suppliers considerably. Often, they offer many different products. In the eyes of the customer, the number of alternatives increases. Brand identity is crucial in mobile Internet. The great number of alternatives and the fact that a customer does not want to think about every purchase give brands an added value. Often customers select brands they are familiar with. Universal brands can use mobile Internet to get directly to the customers, surpassing the channels of distribution; even on busy Saturday mornings at the supermarket. The power of the brand supported by advertising will yield more traffic. Whether traffic is in the store or on the site, does not matter, as long as the result is revenue.

Internet superpowers such as Amazon.com, Yahoo!, and @Home are busy organizing their product line, distribution, and support for the European market. The large amounts that are set aside for e-commerce investments by old-school famous brands set the trend. In the United States, Toys 'R Us and Walmart have launched a counterattack, with their own e-commerce toy sites. Lego, Nintendo, and Mattel offer large sites with entertainment, information, and the opportunity to buy their toys and games. Flagship stores by brand manufacturers are no longer a new phenomenon. The question is if the customer is loyal to the store or to the brand. Much will depend on the extent to which companies draw customers to their sites with differentiated products. Push advertising via SMS can help in this case.

Market research agency IDC takes this one step further. IDC predicts that to make the identification with their brand even stronger, and to stay ahead of the competition, companies such as Yahoo! will give free mobile phones to their customers. The home page for the mobile telephone should create customer loyalty to the brand. Even though Virgin confirmed this theory by giving away One-2-One mobile phones to good customers, some reservations are not out of place. The investment in mobile phones is an assault on return on investment, even for telephony companies. The additional revenue for a company has to be considerable to justify the investment. Decreasing prices for mobile phones and the rise of customized applications can justify this to a certain extent. What is $200 for a mobile phone if it gives you the security that a good customer will conduct all its business with you? It is a sound way to create customer loyalty in sectors in which the profit per customer is high and competition strong. Think of financial service providers, ICT, portals with high revenue per customer, suppliers, or telecom company alliances.

Physical versus Virtual Presence

The importance of physical channels is diminishing more and more for commodity products. The question, of course, is if a virtual store executes the physical delivery correctly. A nice Internet site is important; almost as important as the logistic and IT infrastructure. As soon as trust is broken, however, as goods are not delivered or delivered too late, the chance of repeat purchases is zero and bad publicity is inevitable. If the wrong goods are delivered or the invoice is incorrect, the supplier will not be in the market for long. The fact that two thirds of all Dutch companies answer emails too late or not at all is also a sign that the success of this new media is far from maximal. The barrier to entry for most parties without logistic experience is therefore higher than anticipated.

From the startup period for WAP, you can see that experience on the Internet provides a great advantage to companies that start to offer their services via WAP, especially on the back-office side. The way products are offered is a new phenomenon for all parties involved. The newness of the medium gives early adopters a considerable advantage. User interaction and user interfaces have to be improved step by step. The limitations of the mobile phone push the designer in a framework that forces reduction to the core essentials. Internet experience helps companies get on their way faster, but it is not enough. The acute need for money and predicted shake-out among e-commerce shops underlines that good organization of the front and back office costs more money than new customers will contribute. E-shops that can't handle the peak in orders incur a lot of extra costs, but if the goods are delivered too late (or not at all) customers will not come back to the e-shop.

Deep Pockets

Suppliers for services aimed at consumers have to invest heavily in brand awareness via traditional media to get enough visitors to their Internet site. According to Competitive Media Reporting, the advertising spending of Internet companies in the United States has been minimized since 2000, this while media spending increased from $650 million in 1988 to more than $3 billion in 1999. In most cases, however, it's better to squeeze costs in operations than stop advertising, because without advertising, income will also decrease. For mobile e-commerce, considerable advertising budgets are needed. The simple fact that the screen is smaller than that of a PC makes it more difficult to find companies on the mobile Internet. To get visitors, investments will have to be made in branding. However, money is not the biggest barrier. Despite the fluctuation of the stock markets, going public is still a good alternative for young companies to finance further growth. Otherwise, there are plenty of existing companies that are willing to pay good money for expert advice in this area.

Knowledge

Knowledge and speed are in many cases bigger barriers for new entrants than money. The hype surrounding WAP and mobile Internet pressures companies to enter the market quickly. The available knowledge of the Internet and mobile communication is a bottleneck in realizing a solid architecture, if even possible at all. Disconnecting current systems could result in significant delays. For many companies, the knowledge of customer needs and customer expectations is a much larger problem, especially in determining what to offer to your customers in a mobile environment. Boston Consulting Group research shows that almost half of the e-commerce projects that existing companies start result in a negative margin because projects are started from an IT perspective without taking customer needs or competition into consideration. This is nothing new. The situation is much more threatening for existing companies, as Internet start-ups are able to offer new applications with a shorter time to market. The need for speed and the limited knowledge explains the enthusiasm for mobile Internet among consultants.

Table 7.3 summarizes opportunities and threats of mobile Internet from the perspective of new entries to market.

Table 7.3
Opportunities and Threats from the Perspective of the New Entries to Market

Quickly creating an innovative brand for national niches could be attractive on a global scale.
International players have the means to roll out strong brands worldwide.
Service offerings can be adapted to national or even local preferences for existing and new competitors in a relatively cheap way.
Existing companies are not always aware of the opportunities of mobile Internet for a number of industry branches, to solve problems, and provide an added value to the customer.
Loss of customers as a result of rapid innovations of (new) competitors that result in a breakthrough in lower prices or improved services.
Taking advantage of the strength of a brand to obtain new customers and keep existing customers, especially in industries in which there are plenty of alternatives.
The opportunity to service customers in places and times where a physical location was not possible or profitable.
A beautiful WAP site is not enough. Logistics can be complex and expensive, but remain an integral part of the service. Many companies already have this expertise at their disposal.
For the development and management of a WAP site, Internet experience helps, but is not enough.
Knowledge of customer needs and expectations in a mobile environment is scarce and valuable. Opportunities exist for companies with knowledge in this area.

SUBSTITUTES

A New Communication Channel

The opportunities mobile Internet offers besides traditional communication media are based on lower cost and the direct personal approach. Internet communication spending is only a small fraction of the amount spent on traditional media. For mobile Inter-

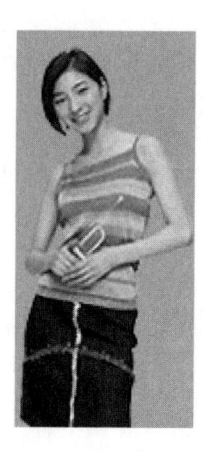

Figure 7.5
The introduction of i-mode in Japan has led to a rise in speech traffic and not to substitution.

net, this will not be different. WAP advertising is already taking place, despite the limited target audience and the limitations in speed. The high attention value and high degree of personalization can help mobile Internet grow into a very valuable media. The effectiveness is easy to measure and the accuracy is out of this world. However, in the area of WAP advertising, much experimenting will have to be done to find the right balance between visibility and user acceptance. In communication between humans, mobile Internet will find a place, but not at the expense of alternative methods of communication. In Finland, the increase of SMS went hand in hand with the increase in the number of telephone calls made. In Japan, the use of i-mode (Figure 7.5) lead to a 15 percent increase in regular phone calls, as people communicated more and more.

Information

A lot of Internet believers predicted that newspapers would become obsolete because people would read the news via the Internet. This hasn't been the case, because people want to read their newspaper in places where you don't bring your computer, like in a taxi or on the couch. With mobile Internet, you can read your newspaper anywhere. Will mobile Internet replace newspapers? The most likely scenario is that mobile Internet as a source of information is an addition to other media, like TV, radio, newspapers, and magazines. People's need for information and communication appears to be limitless and mobile Internet can supply information at times and in places existing media cannot.

Most people will not sit on their couches at night with a remote control for their TV in one hand, a newspaper in the other and then check their mobile phones to read a news column. Substitution of information services does not seem to be an issue in the newspaper industry. It is more likely to be an addition to the existing media. For the younger generations, the new media will be an everyday occurrence and will therefore take up a more prominent place. A shift in information consumption is more likely to take place among young people than among older people. This has been proven with reading the newspaper versus watching the news on television. This means that most newspapers will not have many customers left in 20 years because their current customer base will grow older and die. Younger people are not attracted to the traditional newspaper. They are more accustomed to multiple use of different media: Walkman on, computer on, and watching TV while sending an SMS.

Distribution

In industries where the costs of entry to market are low and the differences between products are small, mobile Internet forms a direct threat as the alternative channel of distribution, even more so if the differences can be made transparent and the service and brand awareness are not strong. A new supplier can offer a product or service much easier and cheaper via mobile Internet. The customer can be reached personally with an offer to his or her specifications. How many suppliers offer this service at this point in time? As soon as a customer bites and switches suppliers he or she is locked in. Winning these customers back is not easy. On the other hand, mobile Internet offers the existing companies in an industry the opportunity to be the first to distinguish themselves by offering improved services to existing customers. This way your customers can look forward to better services and possible entrants to the market are excluded from the start.

The combination of physical outlets and electronic outlets is espoused by the existing retailer. The objective is to get into contact with customers in as many ways as possible, and in the way that customers enjoy. Mobile Internet could play a role in this. If the customer in a sporting goods store considers buying a pair of Nikes, the knowledge that another sporting goods store around the corner has the same shoes on sale can make the customer decide to leave the store. A customer can also decide to order that pair of shoes via the mobile phone at a later date after he or she has tried them on in the store—at the same terms, of course. Mobile Internet can make existing stores more attractive and supplement them as a means of communication and interaction. As part of the multichannel strategy, efficient use of means such as communication, logistics, and payment infrastructure can be made. If mobile payments can be executed safely, simply, and cheaply, it is not unlikely that people will purchase products via the Internet or TV and pay via their mobile phone.

Entertainment

Internet lotteries and games via the Internet are in full swing. Gambling and playing games via the mobile phone is a new offering to users. The user does not have to go to arcades or to casinos to play games. The environments are different, of course. With the introduction of the television, the end of movie theaters was predicted. Obviously the environment and ambiance of a movie theater is enough reason for people to go to the movies. This will not be different for casinos or game arcades.

The arcade games at restaurants will likely be affected. Users will soon be able to order ahead to shorten the waiting time. Even during the wait, users can play games on their own mobile phones. The advantage is that the game can be continued after dinner. It goes without saying that these games also appeal to people who have to wait in places where there are no arcade machines, for example, bus stops, gas stations, waiting rooms, and in the back of the car. Impulse buys such as scratch-off games and lottery tickets will no longer have to be purchased at kiosks. Via the mobile phone, you can buy, pay, and scratch your lottery tickets when you want. You will no longer have to wait for the drawing, as that occurs directly after purchase where and whenever you want.

Transactions

The substitution of small change by payment mechanisms for the mobile phone appears obvious. This application will yield large cost advantages, especially for the vending machine industry. If a mobile transmitter and receiver has been built in to facilitate payment, the status of the vending machine can also be reported. Users do not have to carry around a new card or make a deposit. The application will also accelerate the use. You could also wire people money instantly. Would this put an end to borrowing money from friends? For the near future, total extinction of cash is fiction, but the mobile phone offers so many advantages for all parties involved that payment via mobile phones will take off, with consequences for stores, ticket booths, and other places where cash is used.

Customers will expect to be able to pay via their mobile phones in the future (see Figure 7.6). As indicated previously in this chapter, methods of payment are a necessary condition, but don't create a differentiating competitive advantage. The way new methods of payment are integrated in services could provide a competitive advantage, however. To pay right away on placing an order prevents the user from having to wait to pick up the purchase. Ordering movie tickets via the mobile phone and paying for them at the same time is great, but only if you can walk straight to your seat once you get to the theater. The fact that it is easier for the user to obtain spending reports could be an advantage of paying via the mobile phone.

Table 7.4 summarizes opportunities and threats of mobile Internet from the perspective of substitution.

Figure 7.6
The Sagem MW939-E phone has a special slot for bank and credit cards to facilitate payment.

Table 7.4
Opportunities and Threats for Substitution

Mobile media spending for WAP advertising will remain limited in the next few years, in comparison to traditional media spending.
Mobile Internet as an information channel complements, not substitutes for, traditional media, especially for adults.
Products with low switching cost, small product differences, and little brand identity provide opportunities for substitution via mobile Internet.
As part of a multichannel strategy, mobile Internet cannot only be used as a new channel, but also as support for existing channels.
Entertainment via mobile Internet could take over certain games, such as arcade games and lottery tickets.
Payment transactions via the mobile phone will increase for vending machines, as well as other transaction situations. Paying via the mobile phone is only interesting for facilitators in this area. Integration with the service offers opportunities for many companies.

SUPPLIERS

Direct Sales

Mobile Internet could be an instrument for customers to skip traditional sales channels and to buy direct from the manufacturer. This is no different from the Internet. The impact will be greater as it applies to more product categories and to more instances. It is an attractive alternative for users, especially in areas where the added value of the salesperson is related to the supply of information, location, and product line. If the salesperson does not have answers to certain questions or if the customer feels ill advised, electronic editions of consumer reports can help in making decisions in the store.

Customers want to see, feel, or sometimes even try on products. However, the customer will be less inclined to buy a product if it is cheaper electronically. Dressmart.com already offers the service of buying clothes via the mobile phone. If the item is not really necessary or if the customer would rather have the item delivered at home, the possibilities are even more extensive. Why would you not order those sneakers from Nike directly? Dell has done this successfully for years with its PCs via the Internet. The bankruptcy of Boo.com indicates that e-retailing is not a risk-free business, but large manufacturers have a sounder financial base to absorb a startup loss if the advantages are large enough in the long run.

The concentration of suppliers and the increased importance of the global economy make it attractive for brand suppliers to adapt their channels of distribution. They can contact their customers directly. Global brands communicated in global media have put suppliers of consumer products in a comfortable position. Mobile Internet is the mass media of the near future. The possibilities to make subtle changes in brand awareness are easier and cheaper. The examples for customized production are numerous, from take-out pizza with toppings of choice to personally designed glasses. The necessity of direct contact with the end user becomes more important for brand manufacturers to anticipate customer needs. This globalization also gives manufacturers a presence in countries where local physical presence would not be profitable and importers were previously used.

Rise of Vending Machines

The vending machine business offers many opportunities for manufacturers and wholesalers to skip a step in the channel of distribution. Also, manufacturers of goods that are currently not yet sold from vending machines have opportunities here. In Japan, it is not uncommon to get meals or even shoes out of vending machines. As previously explained, a mobile device will be built in to the vending machine for payment.

Nobody will have to go empty the change out of the vending machine and deposit change at the bank. As no cash is kept in the vending machines, it will be less tempting to break into a vending machine. Vending machines will therefore be placed in more locations where 24-hour availability of products is expected. Besides solving the cash problem, mobile Internet offers one other large advantage: control from a distance. The vending machine can give a signal if there are more products sold than usual or if something is not functioning properly. Restocking vending machines can be controlled more efficiently and maintenance could be done remotely.

Using Extranets

Setting up mobile access to intranets and extranets for various parties in the value chain makes the relationship stronger and offers cost advantages, customer knowledge, and a shorter time to market without suppliers starting up new activities. This way, both sides benefit. The number of business-to-business portals increases and, where relevant, mobile access could be created.

Table 7.5 summarizes opportunities and threats of mobile Internet from the perspective of suppliers.

Table 7.5
Opportunities and Threats from the Suppliers' Perspective

If the added value of the supplier is related to the supply of information, physical location, and the extent of the product line, mobile Internet forms a direct threat.
Supplier concentration and the increased importance of global brands that can supply tailored products have opportunities to skip mediators in the value chain.
Suppliers will decide faster to select the vending machine as the distribution channel and to take care of logistics in-house.
Introduction of extranet/portals with mobile access will improve cooperation with suppliers.

8 WAP in the Real World

In this chapter...

- CMG—ICT Services Group   192
- Emerce—E-Business Publisher   193
- iMedia—Multimedia Design Agency   195
- De Telegraaf—Newspaper Publisher   196
- Scoot—Finder Service   198
- Atos Origin—IT Services Company   199
- Lotus—Software Company   201
- Bruna—Retail Chain   203
- Bibit—Payment Service Provider   205
- 123internet—Content Organizer   206
- AtoBe—Wireless Application Service Provider   208
- Finphone—Financial Information Provider   210
- Siennax—Application Service Provider   211
- Sky Radio—Radio Station   212
- Twigger—Worldwide Email Access   214
- WAPDrive—WAP Portal   215
- XS4ALL—Internet Service Provider   218

The introduction of @info has led to much interest in the Dutch business world. A large number of entrepreneurs started with WAP in 2000. The @info portal started with 30 services and hosted more than 300 sites in 2001. There are also numerous companies that are not connected to @info but have a site for customers or for their own use. To give you an idea of what other entrepreneurs are doing with WAP, we have interviewed a number of them from a range of different companies, from radio stations to consultants, from booksellers to Internet service providers, from newspapers to game developers, and from financial information providers to search engines. We have asked them to indicate what they currently do with WAP, why they started, what their business model looks like, and the way they view the future of mobile Internet.

CMG—ICT SERVICES GROUP

CMG is a leading information and communication technology (ICT) services group that offers business information solutions via consultancy, system development and integration, software applications, and managed services (see Figure 8.1). CMG provides proven knowledge, products, and solutions for the financial, transportation, telecommunications, media, energy, and government markets.

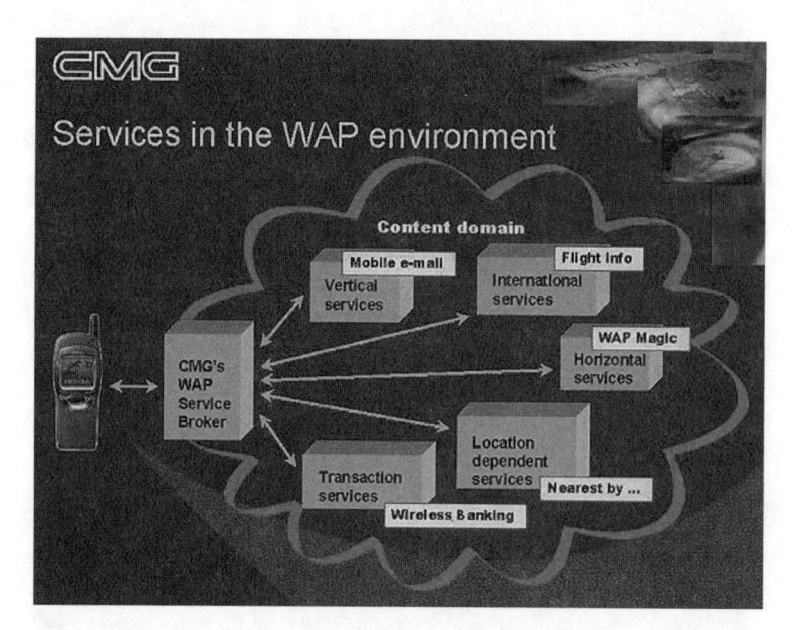

Figure 8.1
CMG services in the WAP environment.

As a worldwide supplier of mobile services and gateway solutions, CMG has been involved in the development of WAP services from the start. CMG has been a member of the WAP Forum since August 1998, and also initiated the development of the WAP Gateway in 1998, which is available commercially under the name WAP Service Broker.

The WAP distribution channel is a strong medium for supplying mobile information and transaction possibilities to users. When WAP is used as an additional channel of distribution in combination with, for example, stores, Internet, and interactive voice response systems, customers can select access to services of a company via the distribution channel of their choice. Critical success factors are connections to back-office systems, 24-hour availability, scalability, and integration with other channels of distribution and security.

CMG has, in cooperation with telecom providers and mobile device suppliers like Nokia, Ericsson, and Motorola, helped to establish the international basis for WAP via the WAP Forum. CMG sells the WAP Service Broker worldwide to well-known mobile operators like Vodafone. CMG has several projects to help clients roll out WAP services.

CMG offers a complete range of services needed to take advantage of modern electronic channels of distribution:

- Design, implementation, and management of Web, WAP, and interactive voice response (IVR) applications
- Server park that can host these applications
- WAP Service Broker for communication with mobile devices (both for WAP and SMS)
- Standard middleware solutions for connectivity with existing information systems
- Design, implementation, and management of the connection to existing information systems
- Design, implementation, and management of new information systems

EMERCE—E-BUSINESS PUBLISHER

Interview with Oscar Kneppers, Publisher

Emerce, the best information source in e-business, is produced and exploited via various media: a monthly magazine, Web site, WAP site, seminars, and trade shows. The target audience of Emerce consists of decision makers and policymakers that work with e-business on a daily basis. Many are managing directors, marketing and com-

Figure 8.2
The Emerce Web site.

munication managers, product managers, and Internet managers. Emerce currently offers a daily news service via emerce.nl with an average of 7 to 10 new messages a day. Readers contribute to the content of Emerce with discussions or announcements. An up-to-date news-providing process linked to the user-generated content has provided Emerce with a leading position in the Dutch market. Emerce will soon set up an online shop (Figure 8.2) from which readers can purchase books, research reports, seminar access, and other e-business products.

Emerce and WAP

"Emerce wants to be there, where the readers and advertisers can expect us. Regarding the innovative character of this industry, we want to be the leader in the market. WAP gave us the opportunity to offer e-business news via this medium as one of the first suppliers in the market. This allowed us to apply our mission; to be the best source in e-business. The Emerce WAP site provides readers with free daily news via @info and other providers. The site has been developed in-house. We will also offer other WAP services such as a commerce portal. The objective is to remain the best source in e-business, also via WAP.

"WAP does not open new markets for Emerce, but provides a better reach within the existing target audience and a selection tool for new target audiences in the existing Dutch e-business market." On the business model used, Kneppers reports: "Out: qualified circulation, free content. In: paid advertising, revenue-sharing partnerships. For Emerce, WAP is currently more a marketing tool than a contributor to the profit margins. But WAP is an important point of focus in the objective to stay the best source in e-business."

Kneppers's advice to media companies considering WAP is: "Publishers have to use it as an additional marketing tool."

iMEDIA—MULTIMEDIA DESIGN AGENCY

Interview with Alexander Zwennes, Managing Director, iMedia

iMedia is a multimedia design agency that develops multimedia products, such as Web sites, intranets, CD-ROMs, multimedia presentations, and kiosks. iMedia is also involved in user interface design. iMedia is active in the B2B market, and works for a number of large and midsized customers such as KPN, Unilever, and PriceWaterhouseCoopers. "iMedia develops Web sites. From relatively simple HTML-based sites to large database-oriented Web sites (for example, *www.cbr.nl*, *www.recycle-exchange.com*). iMedia is also very at home with new technologies such as Flash and Shockwave."

iMedia and WAP

"iMedia first got involved with WAP when KPN Mobile contacted us for developing a promotional CD-ROM for @info. We became very interested, and started thinking of the possibilities for WAP. iMedia recently developed a small database-oriented WAP application for KPN Mobile, with the stock value of KPN Mobile as the topic. We also have our own WAP cam (*mmm.imedia.nl*). This is only for fun, as you can barely see anything on the small black-and-white screen. In the short term, we will develop applications for a new customer we are currently developing a Web site for. Through this Web site the user can place transactions of a financial nature. The customer wants to offer one or more of these transaction possibilities to the user as a WAP experiment.

"Right now the WAP applications are an extension of our Internet activities. In the coming year we will expand our expertise in the area of mobile Internet even further and apply this knowledge to the development of applications for our existing customers. WAP is an addition to the existing services for iMedia, and does not yet have to contribute to the bottom line. So far we have put 200 hours into WAP development. We do all these activities ourselves."

Zwennes's advice to companies that consider WAP development is: "Don't let the hype or the negative criticism get to you. Despite the current limitations of mobile Internet, value-adding services can be developed even with the current state of technology. Do not expect to make millions in the short term."

DE TELEGRAAF—NEWSPAPER PUBLISHER ....

Interview with Johan Looijenga, Managing Director

"Mobillion is a content provider aimed specifically at the development and exploitation of mobile communication—infotainment—and transaction services, using voice response, SMS, and WAP technologies. Together with shareholder *De Telegraaf* (the biggest newspaper in the Netherlands), WAP services are offered for news, sports, and financial news. Mobillion also developed infotainment services in cooperation with media partners, such as SBS Broadcasting (a commercial TV station). Together with these media partners, broadband services will also be developed for future UMTS networks. Shareholder and mother company *De Telegraaf* is very active with regard to the Internet. Specifically for mobile services, Mobillion will also develop activities pertaining to the Internet. In combination with Internet and other electronic media, new services will be offered via the mobile phone" (see Figure 8.3).

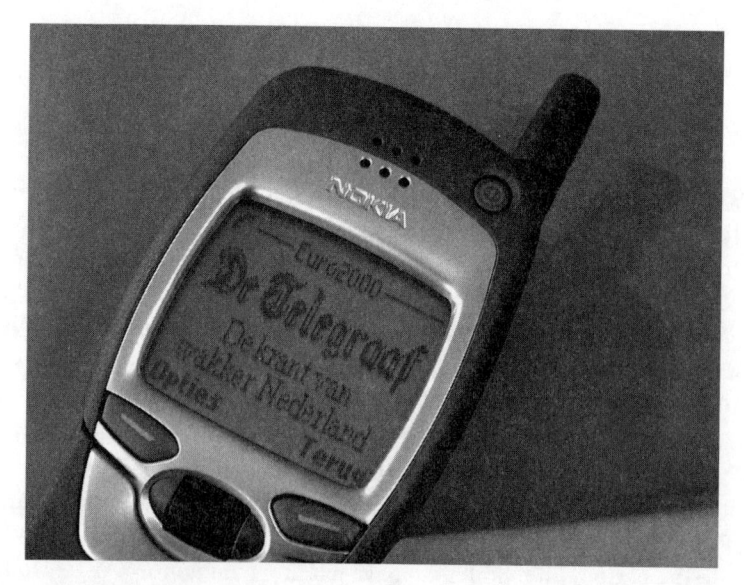

Figure 8.3
De Telegraaf on WAP.

WAP fits into Mobillion's company philosophy: the development and implementation of mobile value-adding services. WAP can be viewed as a new channel of distribution for content. Next to print, Internet, television, and teletext, mobile communications also plays an important role in the expansion of publishing activities by *De Telegraaf* and its associated companies. In the area of WAP, Mobillion offers various content services:

- Telegraaf news link: Up-to-date general news
- Telegraaf sports link: Up-to-date sports news
- Telegraaf weather: Current weather updates
- Telegraaf auto search: Interactive classified advertisements for used cars
- The financial Telegraaf: Financial news and background reports

New services are also developed in the area of infotainment and mobile commerce. "First, offered WAP services are compared to the supply on the Internet. However, the content will become more mobile oriented, when you take into account the personality of the user (Figure 8.4) and the limitations of text and graphics."

Business Model and Objectives

"At this moment, the current business model is based on premium rate revenue from the use of services. This will shift in the future toward the revenue from advertisements and transactions. In the end, a mix will appear from revenue generated by users and revenue generated by advertising and transactions. The objective in the short term is to gain experience with the new medium. In the long run, mobile services based on

Figure 8.4
Mobile content must consider the personality of the user.

WAP will take an important place as one of the most important electronic media via which new publishing concepts, as well as the combinations with other media, can be offered. Publisher *De Telegraaf* found that mobile services offer so many opportunities that they founded a separate subsidiary, Mobillion."

Developing WAP Services

"The efforts in developing WAP services depend on the availability of content. If this content is stored in a database format (for example, XML), and is updated, it is little to no problem to transform the information to a WML format. As this was the case with *De Telegraaf*, only a few days were needed to make the content available via WAP. With regard to the experience gained, new WAP services will be developed."

Advice to Companies Considering WAP

"The use of WAP will have to be part of a wider selection of services via various types of media. WAP can be an added value regarding the Internet or services that are already offered via 1-900 voice response.

"When the expertise on Internet and HTML programming is in house, it is easy to expand this with WAP knowledge and services can be offered more easily."

SCOOT—FINDER SERVICE

Interview with Stein Peters, Product Manager

Scoot supplies categorized information to find companies via all possible media for people from 18 to 49 years old. Internet is an important medium for Scoot. Partner contracts are designed to stimulate Web site traffic, as well as Web advertising and different themes to attract visitors more often.

Stein Peters says, "Scoot is technology driven and wants to offer its information via every possible channel of distribution." Scoot has offered its search service via @info and via WAP.scoot.nl since October 28, 1999. Searches can be made using keywords, topics, or trends. WAP connects seamlessly to the information consumption pattern in today's environments. Scoot has noticed a clear shift in the use of channels of distribution: mobile use has doubled, and the increase in the number of Internet searches indicates a continuation of this trend. When we take this behavioral change into account, WAP is a logical step that links to the demand in the market: information at any time, in any situation, wherever the user may be. Scoot sees these developments as fundamental and an indication of the future. Says Peters, "In the short term, the content supplied via WAP will be expanded. At this moment, WAP is a carbon copy of the Internet. In the future, targeted content will be provided. We want to jump ahead and take maximum advantage of the technology and the continuous adaptation of the content that is suitable for this channel of distribution."

Scoot supplies free information to companies and consumers. If a company is recommended (connected), the company in question pays a lead fee. The WAP service is one of the channels of distribution used. Scoot has created its own WAP site and has invested about 200 hours in content, marketing, and technology adaptations. Peters' advice to companies that consider developing a WAP site is, "Make sure that the data can be accessed easily, so that people can refer back to it, similar to the use of the Internet."

ATOS ORIGIN—IT SERVICES COMPANY

Interview with Evert Hemmers, Senior Consultant

Atos Origin is a company founded on a wide selection of ICT services. Atos Origin's mission statement and strategy reads: "Atos Origin is committed to continue building a unique IT services company, and being the first company to meet many of the demands of the new emerging electronic extended enterprise. In 1999 and beyond, Atos Origin will continue to position its services around Enterprise life cycle management (ELM) solutions." Its customers are large and midsized companies, both national and international.

Within Atos Origin many departments are intensively working on Internet and intranet projects. Atos Origin has its own extensive intranet that forms the basis of the support of the daily workload. The intranet is maintained by Atos Origin, as are the Intranet sites of a number of large companies. Atos Origin is an innovator in the area of Internet and intranet, as the following examples show.

As a unique service, Atos Origin has developed the concept of the intranet hotel. This provides companies with a "collaborative knowledge management environment" for executing projects and pilots, and gives parties an advanced environment that supports the concept of the extended enterprise with a unique combination of workflow, groupware, document, management, news channels, and full-text search facilities. Atos Origin has developed the "aqua browser," a unique concept that personalizes the user interface and enables the user to intuitively use the information and functions.

Atos Origin places value on the expansion of knowledge and capabilities. This can be concluded from the activities in the area of WAP. "First, we do not see WAP as hype. It is true that some analysts view WAP as hype, but Atos Origin does not believe this. WAP, if used for supporting services, transforms those services in a natural way, similar to the possibilities 3G will offer in the near future. WAP is a suitable springboard from the essentials of customer-focused company applications (customer relationship management is a good example) to the mobile Internet."

For Atos Origin, it is important to stay ahead in applying new technologies to its services. WAP is a promising enabler for new applications and thus a logical step. Atos Origin works for large and midsized companies and so it becomes clear that their main focus lies with the B2B possibilities that WAP offers. Gaining experience in setting up intranets using WAP is an important service to customers so that they can gain experience themselves. Atos Origin has set up its employee intranet via WAP in cooperation with KPN Mobile. The experience gained has been translated into solutions offered to the customers. Customers can now make their intranet accessible to their employees. Atos Origin places a high value on customization and supplying a quality product. In the near future, the functionality will be expanded. At the moment, office automation applications have been set up; the next step is implementing ERP applications. Within the telecommunications group, Atos Origin also monitors the developments of WAP and other technologies.

To Atos Origin and its customers, WAP means a new channel of distribution in addition to existing channels. This new channel has limitations. WAP is partly a technological issue (from security to programming WAP sites in WML) for which Atos Origin offers many solutions. Also, in the align phase (brainstorming and creating the correct secondary conditions to m-business) and the operational phase (supplying the infrastructure and ASP), customer support is offered. Atos Origin has a special work group called Esire that specializes in consultancy in the align phase, at which point all their experience will be accumulated. In 2002, a large group of people will be working with m-business. Much will depend on the availability of the supporting techniques such as GPRS and later UMTS.

The research-like WAP project that Atos Origin runs takes up considerable investments. In using this experience Atos Origin can offer customers "onboard possibilities" for WAP at a price of $20,000.

Atos Origin offers the following advice to companies that want to give mobile access to their intranet:

1. Select the applications that can be used for launching WAP (critical "quick decision" information for every location for mobile warriors).
2. Start fast and gain experience inside the company.
3. Design based on performance.
4. Select a scalable approach: Start with a smaller user group, improve, expand, improve, and roll out.
5. The maintenance and support have to adapt to the changing needs of the users.

6. Lower the user threshold to access the well-known portals, as they will develop new and useful functions.
7. For business-to-business applications, security is the most important issue, both organizational and technological.
8. It cannot be predicted what the killer application will be, if there will be one. The only way is trying the function in real life.

LOTUS—SOFTWARE COMPANY

Interview with Tonny van der Greef, Business Development Manager

"Extending the mobile value of Domino to current and new users" is the slogan Tonny van der Greef uses for the software company Lotus. With more than 60 million users worldwide, Lotus Notes has set a record in helping companies increase their productivity and communication in a way that was not deemed possible 10 years ago.

Lotus has opened the "collaborative" Domino infrastructure for various mobile users with different user profiles. Lotus has the vision of "pervasive collaboration, anytime, anywhere, any device," a vision that links to the need in the market. All manufacturers from PDAs and GSM devices launch new WAP-based devices to the market that have the ablility to use data applications and Internet (WAP) services faster and easier. The expected growth is enormous. Looking at the prognosis, the number of mobile devices will soon exceed the number of PCs.

At the moment, WAP offers many new possibilities for mobile users. With the introduction of new technologies such as GPRS and later UMTS, mobile users will receive more bandwidth and thus more functionalities for the more intelligent PDAs. Integration of embedded software for the PDAs is an issue on which Lotus closely cooperates with PDA suppliers such as Nokia and Ericsson.

Market Leader

With the Domino/Notes platform, Lotus is market leader in the area of collaborative work and messaging. With the growth of the popularity of the WAP protocol, the communication possibilities only increase. In addition to existing ways of using a Notes client and/or browser, Lotus is now also able to create a Domino infrastructure accessible via WAP technology and/or SMS technology.

The Lotus Notes client family, comprised of Lotus Notes Release 5, iNotes Release 5, and Mobile Notes, meets the needs of users who want to access information in a way that meets the person, the responsibilities, and the work needs. Domino Release 5 is an integrated messaging and collaboration infrastructure for all these clients. With Domino, IT managers are able to offer consistent functionality for all users, both in the office and on the road. Domino Designer is the basis for application development for the whole client family. This has resulted directly in lower investment costs for training, maintenance, and support for the client software for different user groups.

Mobile Notes brings Notes functionality to mobile devices such as PDAs based on the Palm platform, EPOC appliances, and SmartPhones. Mobile users have the opportunity to create collaborative e-business applications from any location, as Lotus Mobile Services for Domino and Mobile Notes offer access to secure information such as email, calendars, and address books for anytime access to critical information (Figure 8.5). The use of applications that can be accessed via WAP is linked to the standard Mobile Notes licenses. Corporate users can purchase a Mobile Notes license. For hosted solutions, business models focus on the use per month.

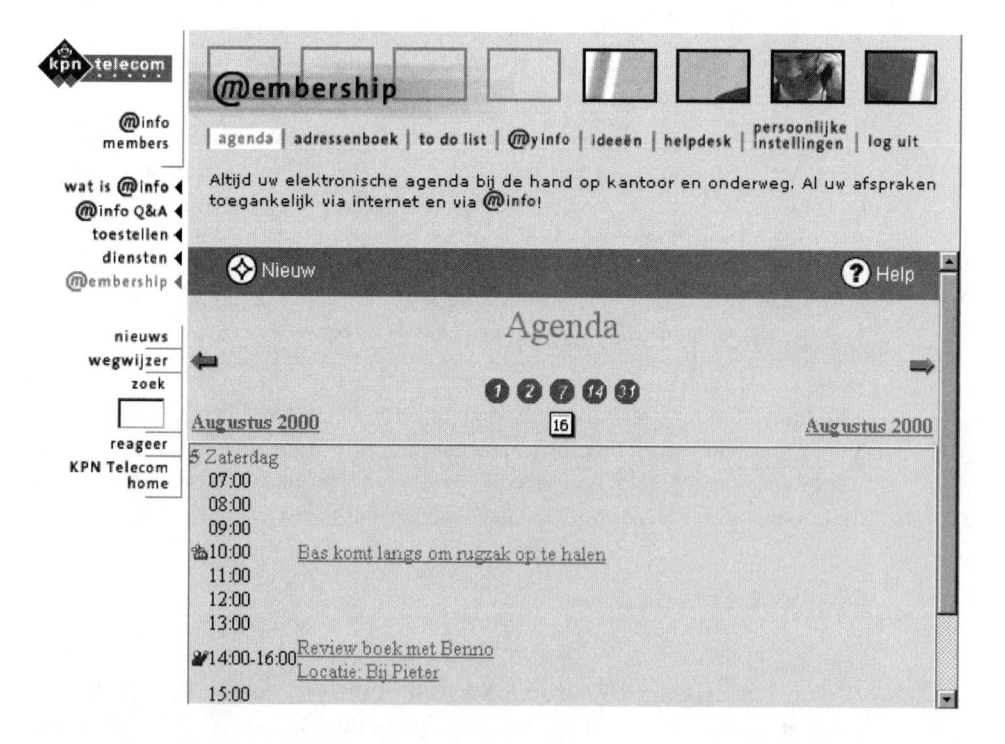

Figure 8.5
The schedule application on @info based on Lotus Notes.

Lotus has also launched Domino Everyplace QuickStart (DEQS) to the market to make a corporate Domino R5 environment accessible to WAP within five days. Together with Lotus Professional Services, it is possible to make the standard R5 mail file (including calendars, to-do lists, and directories) suitable for WAP users. Lotus also introduced Mobile Services for Domino (MSD) 2.0, which makes direct replication possible between PDAs and a hosted Domino environment. MSD uses WAP and SMS over mobile networks such as GSM and GPRS.

@info Info

A real-world application of a combination of Lotus and WAP technology is @info. Users can use various business applications via the display of their mobile telephone based on the Lotus Domino platform, such as a calendar, a to-do list, and an address book. "Together with KPN Mobile, we demonstrate new Domino applications that give more substance to this concept," says van der Greef. "It is our intention to develop more new applications that use the growing number of mobile networks that are suitable for WAP. In the end, all Lotus Domino applications can be accessed via the mobile device. This way we remain parallel to the needs of the customers with our vision of pervasive collaboration, anytime, anywhere, any device."

BRUNA—RETAIL CHAIN

Interview with Michel Koster, Content Manager, info.point

Bruna is a retail chain, active in the market of books, CD-ROMs, magazines, paper and stationary products, cards, newspapers, and computer supplies. In this area, Bruna is the market leader in the Netherlands. At Bruna you can shop for products and services for your personal needs to keep in contact with others, to remain updated, to work from home, and for your personal development. Bruna supplies a current product line, offered in an attractive way. With its chain of stores, Bruna aims to be easy to reach and easy to enter, with friendly and professional personnel.

E-commerce for Bruna means being able to service customers with the correct product and services where, when, and how they wish. E-commerce activities are a good addition to the stores. In almost all cases, Bruna runs promotions simultaneously in the stores and on the Internet. The company has an Internet site (*www.bruna.nl*) and a WAP site (*mmm.bruna.nl*). Since 2000, it has been possible to order a book from Bruna via a WAP phone and pay by credit card. Bruna is the first online store with a WAP site that enables this method of ordering and paying for books. Owners of a WAP phone can order a book at any time and pay using their mobile device (Figure 8.6). Barriers such as business hours are a thing of the past. After reading a book review or seeing the book on

Figure 8.6
Bruna is the first online store with a WAP site.

television, an order can be placed. Gifts are also quickly ordered. Another benefit in comparison to a medium such as the Internet is the lack of log-on time. The book will then be delivered within 2 to 5 working days.

"We have started with WAP, because supplying our services via a mobile device fits in with our multichannel strategy. WAP also opens new markets. Think of the penetration of the many millions of mobile phones and the other devices that will soon be WAP enabled (for example, PDAs). The penetration of the Internet is even lower than that of mobile telephones! The future looks bright for Bruna with the rapid replacement of the existing devices (14 to 18 months) and the introduction of WAP and soon GPRS. The content of our mobile activities is largely comparable to the content of our Internet activities. If we offer the content of a book online via WAP, this content will also be available via the Internet in the desired format. At this moment, we are investigating if personalization is desired and how we are going to integrate this in our media. This way, mobile Internet fits even better in our multichannel strategy. Micro payments via WAP also offer opportunities the Internet does not: certain services to the amount of one dollar independent of time and place."

As Bruna has a flexible back-office at its disposal, the implementation of mobile Internet was a relatively easy task. As all content and software are stored centrally, linking the WAP site to the back-office was simple. WAP now runs alongside of info.point (interactive kiosks that are available in more than 300 stores) and Internet. Keeping the content up to date is very efficient due to the back-office and takes up little time. You enter a book top-10 list once and it appears on all media. Ordering via *mmm.bruna.nl/* is simple. Normally entering name, address, place, and other data via the mobile phone is fairly cumbersome. Bruna has searched for a solution and integration with the Bruna Web site. A customer only has to log on once via the Bruna Web site and can then order via the WAP phone in four easy steps.

1. Surf to *mmm.bruna.nl/*.
2. Select a book and select the option Pay now.
3. Enter the logon ID and password of the Bruna Web site.
4. Enter the credit card number and expiration date. Select Options, Pay now.

Of course, it is also possible to log on directly via WAP. Then the logon procedure consists of entering initials, last name, and zip code. The customer receives a special WAP code to purchase a book faster the next time.

The transfer of credit card numbers that are keyed into the telephone is secure. The security is guaranteed from the WAP telephone, via the mobile network and Internet, to the Bruna WAP server. On the Internet, the certified Hypertext Transfer Protocol with privacy (SSL 3.0) is used for Internet payment security.

BIBIT—PAYMENT SERVICE PROVIDER

Interview with John Caspers, Business Developer

Bibit Internet Payments, founded in 1997, specializes in Internet payments and invoicing. Bibit has more than 500 customers who use their Internet payment services. As of May 2000, Bibit also offers a WAP payment service.

"Our customers for WAP payment service are companies that sell products and services online via WAP. For this they need a flexible and extensive payment solution. Bibit has three years of experience in offering payment services for conducting payments via the Internet. This experience has enabled Bibit to develop a secure payment service (Figure 8.7). The benefits to our customers are that we will constantly update the methods of payment (we already support various credit cards) so that they do not have to take care of that themselves. We also take care of all security issues. We use state-of-the art firewalls and databases and guarantee high levels of payment security. Moreover, we take care of the administration of all transactions for our customers,

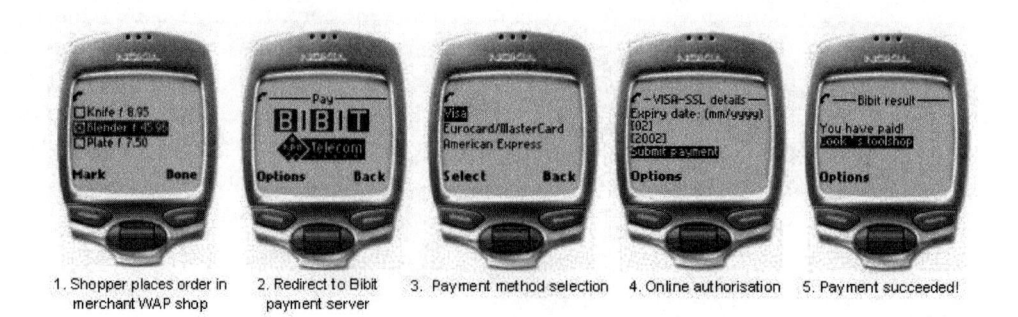

Figure 8.7
Bibit secure payment service.

1. Shopper places order in merchant WAP shop
2. Redirect to Bibit payment server
3. Payment method selection
4. Online authorisation
5. Payment succeeded!

which allows them to know if a transaction was completed successfully or to check the payment history on the site.

"This way, our customers can focus on their core business: the sales of products and services. The reason for Bibit to start with WAP was the fact that it involved the technology of the future, but also because the market wants payment solutions that are suitable for WAP. For our WAP payment service we used the Internet payment service wherever possible. We earn money based on transactions. In the future, we want to expand the number of payment methods for WAP and also the number of WAP payment service customers. We expect that in the future, a large number of payments will take place via the mobile telephone. There will be innovative applications, such as paying via the mobile telephone at the door or in the store. This will lead to an increase in the number of transactions.

"The effort in developing a WAP payment service was relatively limited, as we were able to reuse our previously developed Internet applications. However, developing a WAP site is difficult and there are few people experienced enough. Most developmental work we did ourselves and part of the knowledge we found outside our company. My advice to companies that want to start with WAP: Think of the use for the application."

123INTERNET—CONTENT ORGANIZER

Interview with Wytze Kuijper, CEO

123internet positions itself as a content organizer, offering information and entertainment services independent of the telecom network used (mobile or fixed) and independent of the device (PDA, mobile, and PC). The initial concept of 123internet is to offer a horizontal Web/WAP/SMS portal including integrated services that can be customized to personal needs by the user. The second phase focuses on vertical portals like business portals or travel portals. The first initiative is a youth portal, BaraQuda, launched in nine countries and aimed at high school students. Communication will take place using traditional communication channels that are under the company's control. The target group is fixed and mobile Internet users in a broad sense, requiring easy access to all kinds of information.

The Web portal offers the following categories: News, Financial, Entertainment, On the Road, Lifestyle, Day and Night, Kidz, Shopping, Communication, and Directory Enquiry (Figure 8.8).

123internet and WAP

The mobile phone is more and more accepted, fully enabling one-to-one marketing concepts. 123internet wants to be a mover in using WAP to be leading in development of UMTS. The WAP site *mmm.123wap.nl*—123 Work And Pleasure—is tailor made to individuals and specific target groups. 123internet has built its business based on its 12-year experience with value-added services in Europe and Asia. Services are tailored to the needs of each individual. This Internet business model and services cannot be copied from the Internet to the mobile Internet, because mobile Internet requires a

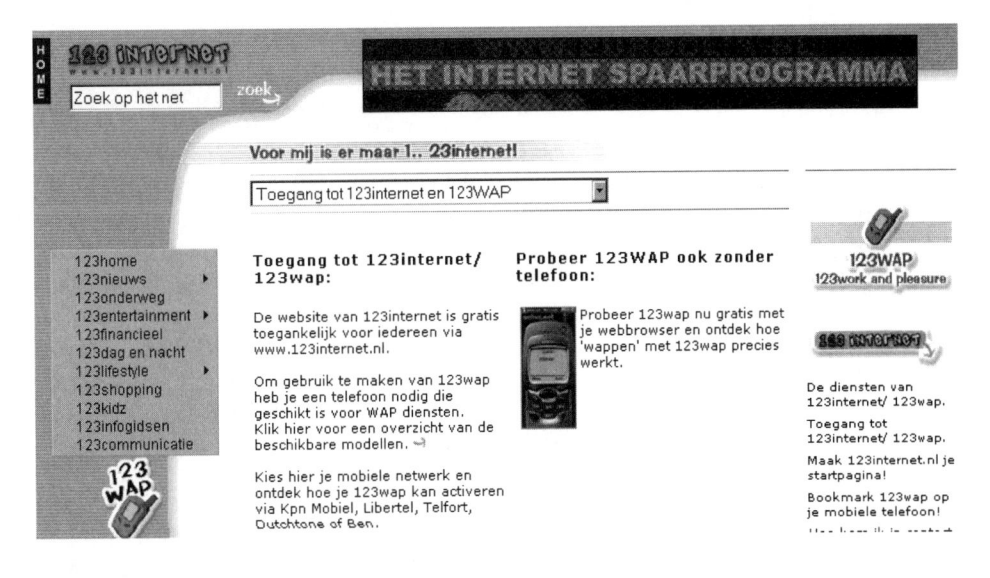

Figure 8.8
The 123internet Web portal.

totally different approach. After finishing the initial learning curve, the new mobile Internet business model will be rolled out geographically.

Both BtoB and BtoC are in the scope of 123internet and mother company FASTWEB ASA. WAP is the initial focus, starting in Norway and opening offices in the other Nordic countries and the Netherlands. Further rollout in Europe, the U.S., and Asia will follow. "Our business model is based on exploitation of information and entertainment services based on Web, WAP, and SMS generating revenues from airtime, advertising, sponsorship, and M-commerce. We aim to serve 10 percent of WAP users this year, growing to 20 percent next year. Development of new service concepts is done by ourselves and in cooperation with technology partners. Hosting is outsourced.

"My advice to companies that start with WAP is to investigate how this new communications channel can add value to the existing channels. WAP can be offered successfully if the back-office is capable of supporting the possibilities of this new medium."

For more information, go to *mmm.123wap.nl* and *www.123internet.nl*.

ATOBE—WIRELESS APPLICATION SERVICE PROVIDER

Interview with Stephan Tieleman, Founder and Arnold Bogaards, Marketing Manager

AtoBe is a wireless application service provider involved with the development and exploitation of mobile multimedia applications. The company aims to support customers in the "mobile arena" with three core activities: consultancy services, application services, and hosting (Figure 8.9).

"It is important to realize that WAP doesn't provide access to the existing Internet services, but that WAP technology enables mobile access to mobile services," says Stephan Tieleman. "These are different applications in addition to the existing Internet applications. The mobile services are specifically aimed at fulfilling a need in the market. Therefore, mobile multimedia services do make up a completely new market," adds Arnold Bogaards. "As soon as a mobile service can fulfill the two aspects of time and place, the user will perceive this as an added value to that mobile service and will be willing to pay for that service. In most cases, mobile services will therefore differ from existing Internet services."

With these activities, AtoBe targets three needs that can be differentiated in the mobile multimedia market.

First, there is a need for companies to make their public information accessible via the mobile phone through third-party portals. Think of ANP news, which offers

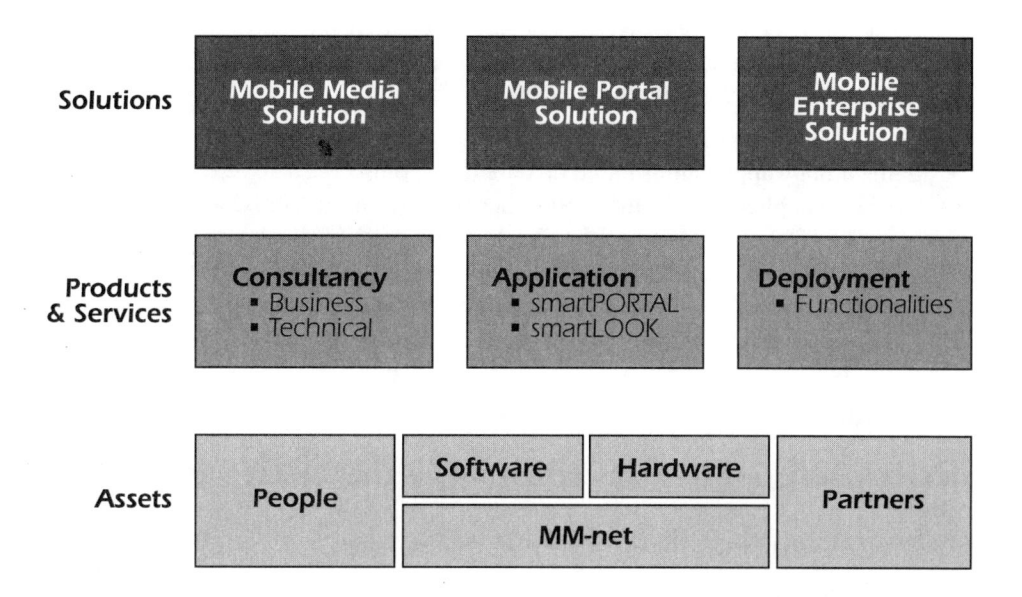

Figure 8.9
AtoBe core activities.

news via the mobile phone within the portal by KPN Mobile. AtoBe offers these companies an end-to-end wireless content publishing solution, based on the three activities mentioned previously.

The second option is that companies manage a mobile portal themselves. Without having to focus on the content of the various services within their portal, they can provide the front office content and also offer third-party content. Think of the mobile operators that wish to expand their regular services, but also existing Internet portals that want to offer a mobile extension to their product line, or even companies with a strong brand name that want to offer several new mobile services, Renault, for example, which provides services for all Renault drivers. For these companies, AtoBe offers the wireless public portal solution.

Finally, there are companies that want to make internal company data available to their employees via the mobile portals to access a product database or the customer database of the email server via mobile telephones. That way, mobile employees can inquire about prices, inventory and customer information, and check their email. For this, AtoBe offers the enterprise portal solution.

In conclusion, it can be said that companies have to ask if they want to exploit a public service. If so, do they want to fulfill the role of portal, and incur high marketing costs? Or do they want to supply mobile information to existing portals? Companies also have to wonder to what extent they can improve effectiveness and customer satisfaction if employees have mobile information at their disposal. Finally, companies need to ask the question to what extent they want to outsource the development of the applications or keep it all in-house.

Based on the customer needs mentioned earlier, AtoBe provides development of the mobile applications based on data from company databases. Next, AtoBe creates a link to applications to the mobile network so that it can be accessed via the mobile phone. The complexity of the development of WAP applications is relatively small, decreasing the development time and cost. The exploitation cost of the service depends on the use of the services and the number of users in the market. As the market for mobile multimedia is still in the early phases, the cost of exploitation is relatively limited.

FINPHONE—FINANCIAL INFORMATION PROVIDER...

Interview with Patricia Broer, Sales Manager

Telekurs Financial is part of the Telekurs Group located in Zurich, Switzerland. Telekurs was founded in 1930 as a service organization of the Swiss banks. The network of ATM machines, transactions, and stock clearing in Switzerland is managed by Telekurs. Collecting financial information, such as stock prices, for the stock trade is conducted by Telekurs Financial. As a result, Telekurs Financial has created a solid position in the market as an international data vendor. The most important customers are banks, but also other financial institutions such as institutional investors and capital managers.

"Our customers (the banks) are actively involved in determining their Internet strategies. This has lead to Telekurs adapting its services. This means that we can now offer solutions based on customer needs. For example, a bank wants to offer its customer a trading facility via the Internet, in which Telekurs can support the bank with stock information for the Internet site. The Internet technology needed is available from Telekurs and various products are operational. Telekurs offers its existing customers primarily financial information, software, and know-how, including WAP functionality."

From the existing infrastructure and databases, an expansion to the WAP protocol is not a large step. In Switzerland, the product Finphone already existed, with which stock rates could be accessed via SMS. In this respect, WAP is the next logical step. Finphone via WAP was introduced in the Netherlands on the launch of KPN Mobile's portal. With Finphone, real-time stock information can be accessed from the Amsterdam Stock Exchange and NASDAQ. The intention is to further expand the product line with the London, New York, and Singapore stock exchanges. Switzerland and the Telekurs office in Singapore will also introduce Finphone via WAP.

The Finphone objective is two fold. On one hand, Telekurs regards Finphone via WAP as a "peripheral product," as it is a business-to-consumer product and a showcase toward traditional customers. On the other hand, the knowledge and experience gained with such services can be used for the benefit of the bank customers of Telekurs. Some traffic is also demanded, so the services are offered at a price per request. Finphone WAP users pay $0.40 per stock quote. The Finphone SMS service in Switzerland receives 50,000 requests per month. For the Finphone via WAP in the Netherlands, this number of queries per month serves as a reference for the objective for the next two years.

"Telekurs had taken over Rolotex in Biel, Switzerland, specializing in Internet applications, two years ago. They have developed Finphone via WAP and take care of maintenance and management.

"When the underlying infrastructure is available (Web server, databases, etc.), creating a service in the WAP protocol is relatively simple. Showing information with WAP can be done within two weeks. Creating a database for registration of use, user access, and CDRs takes up more capacity, depending on the specifications. Telekurs views WAP as an added value to the current facilities and as an extension to the other services. WAP in itself is nothing; WAP is a protocol with which information can be displayed. It is all about offering a service with which it is convenient to access information at any time in any place."

SIENNAX—APPLICATION SERVICE PROVIDER..........

Interview with Michiel Steltman, CIO

Siennax is the first European "pure-play" application service provider (ASP). As a pure-play ASP, Siennax has a clear mission: hosting Web-based business applications and supplying connectivity services to business users. For the customer this means that maintaining a complicated ICT infrastructure is no longer needed, so they can focus on their own core activities.

Siennax creates virtual organizations by offering access to company networks anytime, anywhere. This is possible via a Web browser, but also wireless with, for example, a WAP device. Siennax uses telecom and information technologies for its customers so that they can use the Internet effectively to streamline IT organizations. As an innovator in the ASP market, Siennax offers not only Web access and storage services, but also an extensive portfolio with business applications such as intranet and extranet, e-commerce, Web design, e-learning, knowledge management, document management, project management, and secure email.

Siennax supplies generic hosted applications, both direct and indirect via resellers and distributors. The customers pay a fixed amount per user per month. Siennax also provides more specific, high-end applications, in combination with consulting services. Siennax also offers connectivity services. Says Michiel Steltman, "We believe that when you 'rent out' applications via the Internet, you also have to be able to guarantee access to those applications!"

Siennax and WAP

Siennax wants to offer easy access to company information to their users of the Intranet Suite. Via WAP, various functionalities of the Intranet Suite are accessible. By cooperating with KPN, this service can be accessed via the @info gateway. Moreover, it is an opportunity for Siennax to gain experience in mobile computing in general and in offering services via mobile WAP devices in particular.

At this moment in time, the users of Siennax Intranet Suite have the opportunity to access email, the address book, and company news via a WAP-enabled device. The email function is fairly extensive and offers an addition to the Web mail function of the Intranet Suite. In the near future, more sources of information will be connected to WAP, allowing for better integration with other mobile devices (e.g., PDAs). By making functionality WAP enabled, Siennax offers its customers the option to access company-critical information, whenever and wherever the user wants. Says Steltman, "Our ideal is making our information available via any mobile or handheld Internet enabled devices 'any time, any where.' For this, we depend on the WAP technology available on the market."

The WAP enabling of functionalities for the Siennax Intranet Suite takes place in-house. The WAP-enabling development process took eight months. As advice for companies that consider starting with WAP, Steltman indicates, "Define the application well and focus on retrieval, as the possibilities for data entry are still limited."

SKY RADIO—RADIO STATION

Interview with Rob Korver, CTO

"Sky Radio targets a large audience. The primary audience is those individuals between 13 and 49 years of age. Sky Radio is very active on the Internet, with an extensive Web site. We predict that the Internet will play an important part in the area of radio, and thus the Internet increasingly becomes a more significant part of our activities."

Sky Radio offers an additional service to listeners by offering the current playlist and information on the radio station via the WAP sites. The most important part is the current playlist—Sky Radio plays music nonstop, without intervention of DJs—which allows a listener to search for the music that will be played, or has already been played. Says Rob Korver, "We see WAP as an extension of the Web site. As the technology is not complicated, we were able to complete the transfer of the content from the Web site to the WAP site before the launch of WAP in the Netherlands."

Sky Radio views the WAP service as a value-added service for the listener. The service is free for the user and in the long run, the cost may be covered by e-commerce-like activities and sponsoring. In the long run, the WAP site will become more interactive by using quizzes. Sky Radio regards WAP as an additional media and the cost is carried by the promotion budget. Sky has developed the site in-house. Says Korver, "A standard WAP site such as we have now was not complicated to create. Development was done in-house and took no longer than two days. The link to the playlist is a development of a few days and maintenance of the site takes up about three hours a week.

"At Sky Radio, we always try to take advantage of the latest technologies right away. That we would use WAP for 'something' was only logical. WAP is not exactly rocket science and some knowledge of HTML programming enables the programmer to pick up WML programming rather quickly. These days, almost every Web server is suitable for WML files, so it is easy to host the WAP site on the existing Web server. A separate WAP server is therefore not necessary initially.

"On the first Sky Radio Web site, *www.skyradio.nl*, the playlist pages are the most viewed sites. Listeners appreciate the service and enjoy reading the list before the show airs. This made it a very logical step for us to place this information on our WAP site. When KPN Mobile started with @info, Sky Radio was there with its own WAP site, which can be accessed for free. Besides the playlist we also offer the most current ratings and some general information about the radio station. In the future, the service will be expanded with e-commerce-like activities such as selling CDs or merchandise."

Korver's advice to companies considering starting with WAP is this: "Think about what a WAP user wants to see. A Web site is often created as a front to a company. This will not work for WAP; a WAP user searches for specific information."

TWIGGER—WORLDWIDE EMAIL ACCESS

Interview with Gabriel Troostwijk, Managing Director

Twigger is an Internet service that enables you to use your current email account worldwide using nothing more than a Web browser without having to register or have any knowledge of configurations. You fill out the user name and password and from the Internet provider list, you select your provider. Your email address remains the same so that you only need to have one email address.

"Through one of the producers of the software Twigger is based on, we had been in contact with the predecessors of WAP. This was called HDML and was an American system. When WAP finally did come about, it was not very difficult for us to make our services available via WAP. We came into contact with KPN Mobile, who was working on a portal for WAP users at the time. They were negotiating with several so-called content providers. We had already finished this process and so we waited until the rest was ready, too," says Gabriel Troostwijk.

"WAP connected seamlessly to our services. People already could access their email account via the Web worldwide, but now they could do so using their mobile phone. This meant that you could access your email at any time, anywhere!

"When KPN Mobile contacted us with their plans, an important question arose: Will we let users pay for our service or not? Via the phone it was now possible to charge customers for small amounts. This could change the whole model of free Internet.

"We decided that we would not charge for our service. One of the most important issues of the Twigger service and name is the fact that the user really gets something valuable without asking something in return. Our service is free and the user does not need to register. The users also do not receive unwanted email from our other companies or us. Twigger is fun and easy to use. And so we had to apply this model to WAP. We did not communicate the 'free of charge' issue to our customers, as we still had not figured out how to make money with this extra service. Via the Web, we were able to generate revenue via banner advertising.

"The KPN Mobile service started and we did as well. A few months later, I was talking to colleagues from .bone (an interactive advertising agency) at a Monsterboard party. I had had the idea of WAP advertising for a while now, so that we could continue to provide our WAP services for free. I discussed this with the folk at .bone, who worked for a company called XOIP. XOIP might be interested in this. If we would do this, we would be the first to have a WAP advertisement in the Netherlands and the second worldwide. In three days, everything was settled. We went live with an advertisement from XOIP and Emerce. Emerce was so enthusiastic when I told them the secret of what we were about to do that they to wanted to be part of this. And so we had additional benefits from WAP. We had a service that was completely in sync with our Web service, from which we could generate extra publicity and revenue. Later, KPN Mobile became one of the WAP advertisers on our site.

"At this moment, WAP is surrounded with much negativity and people believe it will not amount to much. People say it is too complex and slow. I believe WAP has a solid future. However, people have to first realize what WAP really is. It is believed that only a few phones use WAP. It won't take long until people will travel with more advanced devices, comparable to PCs. These PCs will have to be able to communicate with the Internet. GPRS and UMTS are not the successors of WAP, but enabling technologies that make WAP faster. The only danger lies in the fact that other similar protocols will be developed soon that will be faster than WAP.

"Mobile Internet, which is mainly WAP right now, has a big future. WAP is not necessarily an 'in between' product, even though it will have to be improved continuously. I believe that the future looks bright, and there is no need for negativism. As long as you do not say no, you are probably right!"

WAPDRIVE—WAP PORTAL

Interview with Andreas Birnik, Director, Wireless Products

WAPDrive (*www.wapdrive.net*) is a prominent WAP portal with a number of personal mobile Internet applications and information. WAPDrive is used in various media instead of just via WAP. "We expect that the majority of our users use a PC to access WAPDrive applications and information when they are at home, and their WAP phones when they are on the road," says Andreas Birnik.

WAPDrive is a portal by FortuneCity, a new media network with 24 Internet sites in nine different languages. It is a network of community and vertical information sites in the area of games, sports, and music. FortuneCity offers its users 100 MB of free storage space on the Web and tools to create a personal Web site. Based on this, expansion of the business model to WAP is a logical step. "We started an internal discussion on WAP in the middle of 1999. The real work started in October and we launched WAPDrive in March 2000. WAP is 'funky' and helps to position FortuneCity as one of the most innovative Internet companies. Our strategy is to become platform independent and so give our users access to the information and applications of a number of platforms such as PCs, Web TVs, WAP phones, and PDAs. Winning companies will offer a wide range of quality information and applications to users in the future that are platform independent."

WAPDrive Service Categories

The personal publishing tools in WAPDrive offer both a simple and an advanced option to constructing a personal WAP site.

Users without programming skills can build a WAP site in a few minutes complete with text, graphics, and links. The WAPDrive WAPtor is a WML editor for Windows, with which users can create a WAP site. We believe that WAP sites created by WAPDrive users will be kept up-to-date and many local sites will be created that focus on a specific geographical area. This creates the foundation of future location-dependent services, company guides, and mobile commerce. The user-generated content has the big advantage of being free for WAPDrive.

Applications

The application part focuses on an address book, calendar, and email. The calendar and address book can be synchronized with Microsoft Outlook. The next step will be to integrate with the applications for new functionalities. Examples of this are email answer functions to the owner of a WAP site or the possibility for a user to publish his or her calendar via a Web site.

WAP Portal

WAP-related content is an important part of WAPDrive. Current functionality includes a WAP emulator by Gelon, a WAP directory by Gelon, a search engine for WAP sites, reviews of mobile phones, a list of different WAP gateways, WAP news, links to other WAP-related sites, and "cool sites" built by WAPDrive users.

For WAP telephones, the WAPDrive portal offers the following options:

- My WAPDrive: Leads the user to calendar, address book, and email
- Partners: Links to WAPDrive's partner sites
- Directory: Gelon's WAP site directory
- Search user sites: A search engine to find sites created by users
- Join WAPDrive
- Get Information
- FAQ

This portal (see Figure 8.10) can be found at *www.wapdrive.net* (from a WAP-enabled phone).

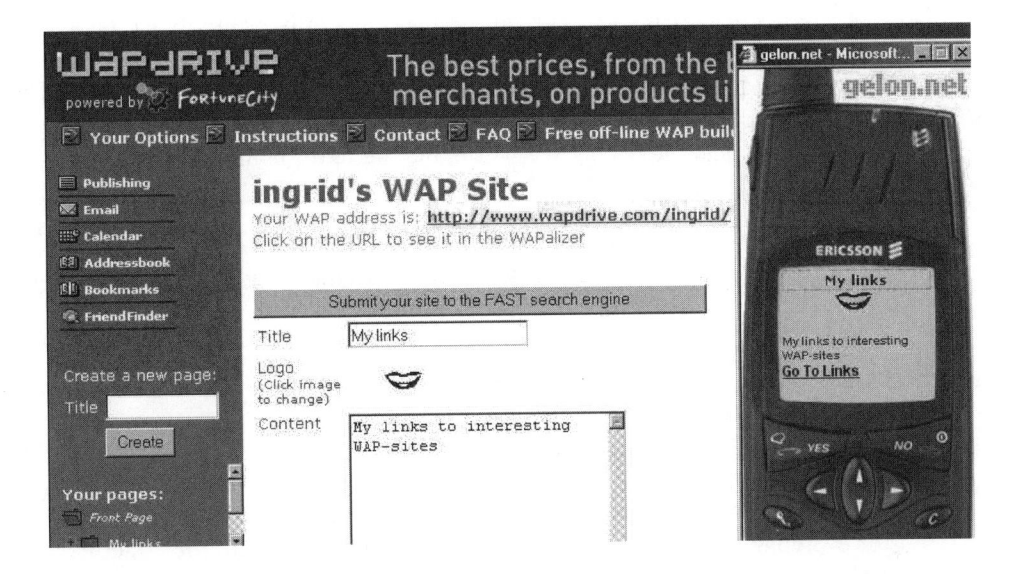

Figure 8.10
WAPDrive WAP homepage builder.

Information Partners

The information part is currently aimed at mobile games. POPEX (*www.popex.com*) is a free music stock exchange game, in which the player can trade stocks in famous pop stars. WSX (*www.wsx.com*) is a fantasy soccer stock exchange game, in which players can trade stocks in soccer players, teams, and coaches. The forthcoming sites of other parties will offer bets, golf information, and funny psychometric tests.

Target Market and Business Model

Says Birnik, "The target market consists of all individuals that own a WAP-enabled device. Forrester expects that there will be 219 million European users with Internet devices in 2004, of which 132 million will regularly use mobile Internet services. The success in Japan with i-mode indicates that the appeal of wireless Internet access is universal and not limited to Europe. WAPDrive's strategy is to adapt and launch services for new mobile Internet standards as soon as they are available. WAPDrive believes that the ultimate potential will equal the number of mobile Internet terminals in the world. As access to distribution is key to most consumer service products, WAPDrive is negotiating with the most important European mobile operators to include WAPDrive services into their product line.

"For a new media network our business model is mainly based on advertising. Revenue will also be generated from partnerships with mobile operators. As WAP-Drive is building a large WAP customer database, future revenue could come from M-commerce opportunities. Our strategy is to become one of the most important players and offer new and interesting services in the future for both WAP and the new mobile Internet standards."

What is your advice for companies that want to start implementing WAP?

"The faster you start, the better it is. It takes time to get where you want to be."

XS4ALL—INTERNET SERVICE PROVIDER

Interview with XS4ALL's WAP doctor

"XS4ALL (Access For All) wants to make WAP available for everybody. XS4ALL was the first in the Netherlands to offer Internet access in 1993, and now offers all subscribers free access to WAP, via its own WAP gateway.

"XS4ALL has chosen free access to all possible information and therefore does not offer its own, restricted information services. Customers have access to all WAP sites on the Internet via the gateway. The WAP gateway is placed right behind the dial in network, and therefore connected directly to the fast national and international connections of XS4ALL. This direct connection prevents the long connection times that are very common with the demonstration gateways by GSM manufacturers such as Nokia and Ericsson in Finland and Sweden.

A collection of WAP sources can be found at the XS4ALL WAP site, www.xs4all.nl/wap/. Every XS4ALL subscriber can design a start menu or add a personal WAP site to the list of links via this site fairly easily. Every XS4ALL subscriber has unlimited WAP access to the mobile Internet and is not dependent on the services offered by the mobile operator.

All customers can place their WAP site (.wml files) on one of the XS4ALL servers. Business customers can also place their own server in XS4ALL's system space to offer WAP services. This way they are sure of optimal services. With invasion of privacy in mind, XS4ALL has selected software that does not pass on number information. XS4ALL leaves it up to the information supplier whether or not to charge for its services. Some suppliers may have a need to mail invoices for services provided. In that case, XS4ALL will give its customers the option of whether or not to pass this information on. More information on this topic can be found at the Dutch site www.xs4all.nl/wap/. What do we do with WAP? What does WAP cost at this point in time? What can you do with it? Who uses it? Which mobile phones are WAP enabled?

XS4ALL believes WAP is the future and offers its subscribers, the customers, the opportunity to experiment with the technology. A dedicated Web site, a WAP panel, and a WAP doctor examine WAP—in the Dutch language. The WAP doctor is available to answer the question: Does the problem lie with the user, the WAP phone, or the WAP site?

How about the setup options of the different phones? Every GSM manufacturer tries to reinvent the wheel. All interfaces are different. And not only that, but the mini-browsers, used to scroll through the WAP sites using the WAP phone, are also different. Nokia and Ericsson created their own browsers; others have purchased licenses from Openwave. The WAP doctor clarifies this chaos by outlining the different setup menus for all the different phones, and by compiling the FAQs and providing this information to the XS4ALL customers. It remains to be seen if WAP will be the ultimate wireless Internet, but XS4ALL gives its customers opportunities to experiment.

Email the WAP doctor at *wapdoc@xs4all.nl*.

9 Five Steps to a Successful WAP Site

In this chapter...

- Step 1: A Clear Objective   222
- Step 2: The Marketing Mix   225
- Step 3: Financial Analysis   240
- Step 4: Plan of Action   245
- Step 5: Control and Reporting 245

In the previous chapters, we discussed the influence the mobile Internet has on the competitive ability of your company and which companies are already involved in WAP. You probably have your own ideas on the use of mobile Internet for your company. If so, this chapter can help you in realizing your ideas. How do you achieve a mobile Internet site to take the first step in the area of mobile Internet? In this chapter, we describe a five-step program that will contribute to your success on the mobile Internet.

STEP 1: A CLEAR OBJECTIVE

Many companies have started with Internet in the past without a clear objective. An enthusiastic employee built a site as a hobby or a site was constructed because everybody else did it, too. The moment the site went live, the job was done and regular work was picked up again. This is really too bad, as the new medium did not get the attention and the budget it deserved. Especially large companies have the urge to belittle new media at first. Who needs it? There are too few users! Only later do they finally realize that a great number of new companies have entered the market, or that the competition all of a sudden is able to use the new media to dramatically change the playing field; for example, in offering a customizable product at a lower price or faster.

To formulate a good objective, we start with the core proposition of mobile Internet: With mobile Internet, your customer always has access to your services regardless of location in an actualized, personalized, and location-based manner.

The first question is for which target market would this be relevant? Your existing customers or your employees? Could this help you create a new target market for your services that you could not reach with your existing products or services? If you want to give your employees mobile Internet access, you will have to examine how you can improve your services, which customers benefit, and what those customers will experience in reality. It is beneficial to use Ansoff's growth strategies (see Table 9.1).

Table 9.1 Ansoff's Growth Strategies

Market	Existing Product	New Product
Existing market	Market penetration	Product development
New market	Market development	Diversification

Imagine you run an employment agency that helps employers find temporary employees and helps people find temporary jobs. By advertising jobs via your WAP site, you reach a larger audience and you can achieve a higher level of market penetration. Even in areas where you do not yet have an office, you could reach people looking for work. Market development is in the cards for you. To which target audience is WAP interesting? People looking for work will want to use your WAP service. Students, IT professionals, and people with a higher level of education will adopt your service faster than production personnel. Employers will prefer to call, unless their industry requires that they access your services when they are outside of the office at restaurants or events.

You could also work on product development. You could service employers that need workers right now. People looking for work that have several hours available on a particular day to work cannot be placed at a moment's notice. For unschooled labor, a planner could be established on a WAP site, indicating the location, the hours, and the number of people needed to do the job. People looking for work could register for work and the number of hours they are willing to work. This could prove to be an attractive new product for students who want to make some extra money. A pool of available personnel can be made accessible to employers via WAP and the Internet. If such a new product is launched to the market, we speak of diversification.

If you have determined the target audience, it is sound to find out to what extent this target audience can be reached with WAP in the next few years. WAP services aimed at younger people (generally men) with a high level of education will be more successful than those WAP services aimed at older people with a lower level of education. In some cases, you could influence the use of your WAP service within your target market. In the previously mentioned example, students looking to make some extra money could be given a WAP phone. This could stimulate the recruitment of temporary workers and the use of these services for the benefit of the temporary workers and employers. Among your own employees, you also have more influence on your target market. You could make WAP services a part of your employees' daily routine and stimulate this with education. We expect that the resistance of your employees to work with WAP will be much lower than was the case with the PC. Employees are often used to a mobile phone and the use of WAP sites is much simpler than using computer programs for someone who has never worked on a PC.

Once you have decided on your target audience and your growth strategy, then it is important to ask what added value does it offer to the customer to contact your service. This added value can generally be expressed in ease, enjoyment, and profit. Does this make it easier for your customer to purchase products and services? Does it make life more enjoyable? Does the consumer save time or money? This means that you have to anticipate one of the following situations:

- The user wants to be well informed of up-to-date information. Examples include news, traffic jams, airport delays, stock prices, sports scores, new email messages, and so on.

- The user wants to search for something. Examples include telephone numbers, restaurants, ATMs, directions, checking status, and train schedules. Even customer self-care falls into this category, like checking the status of an order or reading FAQs.
- The user wants to order something or make a reservation. Examples are concert tickets, a hotel or restaurant reservation, making changes to airline reservations, and so on.
- The user wants to make effective use of waiting time. Examples are reading emails, checking calendars, or entering business card information into a database.
- The user has to wait somewhere (train or airport) or is bored and wants entertainment (and still come across as a professional). Examples are games, weather reports, humorous news, or erotic images.

With this you can come to a core proposition for your service for mobile Internet. Here are a couple of examples:

- Young people with little time to shop can easily order books anywhere.
- Customers always get access to up-to-date information.
- Customers can see the status of orders anytime, anywhere.
- Travelers will never have to be bored (games).
- Motorists and pedestrians can always find their way.
- Young professionals can always communicate where and when they want.

If your employees function as a customer of your WAP service, the following could be the core propositions:

- Salespeople are updated with the latest information on prices, stock levels, and so on, so that they can inform customers properly.
- Salespeople can check the status of orders when they are with the customer, so that they can answer all the customer's questions related to orders in process, availability, and delivery times.
- Field workers can read their assignments and fill out time sheets and reports so that they do not have to make the trip to the office.

When you have determined the core proposition of your service for your customer, the next question will be: What is in it for you to offer this service? Do you expect to gain new customers? Do you expect that your existing customers will be more satisfied with this service and will order more, or that you will get to know your customers better? That your employees will become more efficient, that your administration will take up less time? Can you save on expenses? Is it good for the company image? Are you ahead of the competition? Will it lead to customer loyalty?

The combination of the core proposition with the expected advantages for the company form the objective. This objective will be detailed and made more specific in the next steps. The following are two example objectives:

With our WAP service, our customers can always obtain status reports on their stock portfolios, allowing them to make more transactions and to become more satisfied with our services.

With our WAP service, mobile users can always search for the nearest location serving beer, resulting in higher traffic for us.

STEP 2: THE MARKETING MIX

When the objective is described, it is time to take Step 2: the application of the five Ps: product, price, place, promotion, and personnel.

Product

From the objective we can create a product description. We advise setting up the product description as broadly as possible and then determining the ultimate service. Later, several things will be deleted or incorporated in parts, but that is okay. It would be a waste to be limited by technological impossibilities from the start, as that hinders the creative process of product definition. Moreover, there are many existing packages on the Internet that are hardly used on the Internet and WAP. Think of personalization tools for creating a personal environment for visitors of a site. This software could also be copied to WAP with limited adaptations. We list here a number of questions that will have to be answered to come to a proper product definition:

- What are the problems you can solve or possibilities you can create for your target audience and customers?
- Is a personalized approach needed or demanded?
- How can we protect the customer's privacy?

- Is this a supplement to existing services via the store, telephone, or Internet or is this a completely new service?
- Does the location of the customer influence the service?
- How up-to-date does the information need to be?
- What do my potential customers find most important to know on short notice, and what is less important?

What are the problems that you can solve or the possibilities you can create for your target market and customers?

In previous chapters, we have highlighted the different trends and possibilities for which we can use mobile Internet. For a sound product definition it is essential to have an idea of how you can offer your customers more benefits with mobile Internet. Think of using services or accessing information regardless of time and place. A kiosk, for example, has a complete portfolio of lottery products. If a customer realizes he or she wants to buy a lottery ticket at midnight, he or she might not be able to get a ticket before the drawing. Mobile Internet would allow users to buy a ticket regardless of the time of day. The benefit to the customer is that he or she does not have to go to the store and could therefore still participate at the last possible moment. If the service is meant for the existing customer, information is often needed to supply the service properly. Often, there is information available on existing customers that can be used in supplying the service. If the service has to provide the customer information on account balances, stock portfolios or the status of an order, this information is needed to service the customer better. With existing customers, it is therefore important to gain insight about the information already available on the customers and to analyze if this information is sufficient to solve their problems or to offer a new service. It is also important to examine the quality of the information. Name and address information is always stored but seldom updated. The customer could have moved and not have a reason to pass this information on. Some WAP service providers, such as @info by KPN Mobile, pass on a unique customer ID number to the information provider. By linking this to the customer number or other information, you can always identify the customer right away. This can be done by entering a customer's account number or another unique identifier when they first contact you. If your service connects to a portal that offers unique identification of customers, you do not have to ask for full customer details each time the person wants to use your service, as you recognize the customer right away.

Data on existing or new customers can be collected via the WAP service, but also via the Internet. Let the customer enter (if needed) a login name and password, which is as short as possible. It is not convenient to enter long text messages on the mobile phone.

Example: Stock Portfolio

> Objective: With our WAP service, our investing customers can always check the status of their portfolio, which will increase the number of transactions and the level of satisfaction with our services.
>
> From the objective we can conclude that this example concerns existing customers. These customers are uniquely recognizable within the bank, thanks to the number of their stock accounts. To recognize the customer when he or she enters the WAP site, a password can be requested as well as an account number. If the WAP gateway service to which the service is connected offers a unique customer identification number, the stock account number will only have to be requested once and the unique customer identification number can be linked to the stock account number to make sure that the customer can be recognized in the future. Next, the customer could select a short access code, allowing him or her to access the stock portfolio faster in the future.

Personal Approach or Not?

A personal approach is not always necessary. There are many applications for which a generic service suffices, for example, news messages. The information is the same for everybody and there is no information on the customers that already use these news facts via different channels.

However, it could be worthwhile to select a personal approach, especially if you expect repeat purchases of your service (see Figure 9.1). If you offer a weather report service, your customers will want to see the weather report for their city on a regular basis. In presenting the weather report of the city shown the last time as a first choice, the customer can access the information much faster than if he or she has to enter the same city name again and again.

Figure 9.1
myCNN.com offers users the option to set news preferences.

For a personal approach, not all customer information has to be available. The data can also be collected on the customers that use WAP services. A personalized service can be offered in the following ways:

- Personalization based on previous behavior

    Example: The customer has visited a restaurant in New York. New York will now already be set as the city, or in the menu, New York will be on the top line and other cities on the second line. If the customer searches for New York again, he or she has to go through fewer selections. Also, the customer had ordered a book by John Grisham last time. He or she will therefore receive a personalized ad for the latest John Grisham book when visiting the site again.

- Personalization based on geographical location

    Example: The service can be focused on the location, where the customer is at that point in time, for services such as "Where is the nearest restaurant?" or "Where is the nearest supermarket?" The location information can be given by the network or asked from the customer.

Step 2: The Marketing Mix

- Personalization based on customer preferences

  Example: The customer has indicated via the Internet an interest in baseball, especially the New York Yankees. When he or she is looking for scores, the baseball scores will be displayed first, with the Yankees' results on the first line.

- Personalization based on interpretation of the data supplied by the user

  Example: Say that a customer indicated gender. Based on that the greeting can be adapted (Mr., Mrs.). Analysis of customer behavior may also have shown that men are interested in a different part of the product line than women. In this case, different information can be presented to men. The customer can also select the language.

- Personalization based on detailed customer files

  Example: You present the customer the value of his or her stock portfolio first.

For personalization, it is therefore very important to uniquely recognize every user that reaches your service. You can always ask the identification number, but this causes a higher threshold to use your service, so connecting to a portal that offers unique customer identification is a great benefit to the user. Location information from the mobile network is a benefit, but this data can also be asked from the user if that is relevant for the service. A zip code is easy to enter, and making a selection from a list of cities is not too bad either. The effectiveness of a zip code as an identifier of the location differs a lot from country to country. In the Netherlands, a zip code will give you the location on the level of a street, but in France it indicates only a part of a town or even a city or rural area.

In the case of personalization, the laws concerning privacy have to be taken into account. What customer data can we store, how do they have to be stored, and how are you allowed to use this data? In the new European Law Protecting Privacy, companies are not allowed to store customer information unless the customer grants permission.

In the media, news appears regularly on the lack of privacy on the Internet. ISPs have especially taken a hit. The registration counsel found that a number of ISPs misuse personal data, for example, by selling traffic data—surfing behavior—without permission. ISPs customers receive unwanted email from companies they don't know with special offers. Consumer Affairs thinks there may be claims for compensation. It is very important for companies to handle customer information with care. The chance is small that a customer that receives special-offer SMS messages all the time would want to do business with you when he or she finds out you violate his or her privacy with spam. Mobile users are probably even less tolerant with regards to unsolicited messages than fixed Internet users are to unwanted email. Receiving unwanted messages via the mobile phone or even receiving calls will be accepted by only a few customers. All of the aspects that make mobile Internet so powerful decrease tolerance: personal, direct, always, and everywhere. Limited privacy on the Internet is a barrier to online shopping. A medium such as mobile Internet can become very successful for online shopping. This means that privacy of the users has to be guaranteed.

At this stage of the mobile Internet it is difficult to say what the limits are, but we can make some recommendations:

- Write up a privacy statement, that indicates what data is collected and for what purpose, and for which purpose it will be used. Clearly communicate this privacy statement to your customers the moment you make yourself known. For example, "We collect your name, address, and city so that you do not have to enter this information every time you make a purchase. We only use this information for delivering orders."
- Provide opt-in and opt-out possibilities; in other words, give the customer the opportunity to indicate what service he or she wants and to change his or her mind later. For example, customers can access a database of houses for sale via WAP and sign up for an SMS service through which the customer receives an SMS message when a house that meets his or her criteria comes on the market. The moment the customer buys a house, or is no longer interested in the service, he or she has to be able to opt out of the service.
- Give customers the opportunity to review information collected about them and allow them to make changes.

Book and CD reseller Bertelsman Online (BOL) CEO, Rick van Boekel, stated in *Intermediair* that, "We would be crazy to share our customer information with others. The value of our company depends on the customer base. Sharing the information will decrease our value."

On the introduction of location determination, much attention will have to be paid to privacy. With commercial services, customers could be warned that their location could be used, and they should have the option to refuse. This can be compared to the way callers are warned about extra expenses when they dial a surcharge number. Storage of the user's locations should not be allowed, unless you have received permission for it from the user or it is so important that it overrides the violation of privacy. Again the measures taken to protect the users are important. Think of communication, opt-in and opt-out, and clarity for the user.

Supplemental or New Services?

If the service supplements existing services, it is important to integrate it with the existing service as well as possible. If you give your customers the opportunity to order products via direct sales, and you now offer this also via WAP, you will have to integrate services. This means that the salesperson has to have access to all customer orders, not only those made via direct sales. The customer will not want to enter details every time he or she accesses the WAP site. Instead, customers will want the existing agreements to also be valid via WAP.

With supplemental services, it is interesting to research the opportunities of triggers. You may want to make your customers aware of the new collection or of the new goods that just came into stock. If customers give permission for this, you could notify them with a message via SMS. With the WAP service, the customers can better inform themselves of the situation. In your WAP service the opportunity could be offered for contact via the telephone.

It is important to keep in mind if the promise you fulfill with your product can be fulfilled faster, better, or cheaper via mobile Internet. Think of the situations that lead to the need for up-to-date information for your customer, the need for a quick search on simple topics, the added value of being able to order or make a reservation at all times. Depending on your business, you may want to take advantage of the wait and commuting time of your target audience, including boring moments.

Location-Dependent or Not?

To what extent is the location of the user important to the service you provide? If you want to guide the customer to the nearest store or bar, the importance of customer location becomes clear. For traffic and weather information, the importance of location is evident. Also, other features like the standard language or language options can be adapted depending on the location.

Current or Not?

How current will the information have to be that you offer? If the user of your WAP service makes decisions right away based on your information, this information has to be up to date. Examples include stock quotes, weather forecasts, and traffic information. When the customer places an order, the information on availability will also have to be current. When the customer books a flight that leaves a couple of hours later via WAP, the flight cannot be full when the customer arrives at the airport. The expectation of current information is very high among mobile customers. This demands much of your systems and processes. If you do not meet these demands, you should question if you should offer WAP services in the first place.

Is Order Important?

In developing a service we have a tendency to be as complete as possible. With WAP it is more important to leave things out. Less is more when it comes to entering information to get to the end result. Find out what the question is and how you can offer the answer or the solution as fast as possible. Anything that does not contribute to this fast solution has to be omitted. If you want to offer the customer the ability to order flowers via WAP, you are better off offering three selections, then letting the customer pick from 100 bouquets. If you want to offer a wider product range to the customer, the following structure can be used: $15, $20, $30, each with three selections. If you design your service it is often easier to start with the basic service and make a plan for expansion.

Price

To create a WAP site, you will have to incur costs. These costs can be earned back in different ways:

- The customer pays for use of your service.
- The customer purchases products or services.
- Advertisers pay to place advertisements on your site.
- The portal or a company pays for your information or application.

For the end user, it is advisable to make the sites on which products and services can be purchased or reservations can be made free. This way the end user does not have a barrier to access the site. From the margin on the products or services, the cost of the WAP sites can be earned back. In general, end users will only want to pay for access to information or applications if they can benefit from it or there are no cheaper alternatives available with the same effort. If the WAP site provides them with access to information that is not easily obtained, a price premium is justified. Think of a bank service that can provide the status of a foreign payment in real time. The same applies to services that facilitate the user. The numerous search and find services on the Internet are an example of this. The question is who will pay the premium: the user or the provider of the services? In the world of telephony there are both toll-free numbers or numbers with an additional charge. On the Internet most services are free. There are, however, some services that are paid services. The payment can be made via credit card, where you receive a password to access the service. There are also some providers that require an additional charge per minute for a service, for example, by switching the call from the ordinary dial-up link with the ISP to a special telephone number with an extra charge per minute. That way the service is paid for via the usual telephone bill.

Services that provide a diversion do not have to be free, such as arcades and slot machines. In determining the price it is advisable to test what price the potential consumer is willing to pay. Not every portal offers the opportunity to charge a price for the service.

You can opt to present your WAP service to your existing customers as an extension of your product line. You provide your customers with more service for the same amount of money or you reward your most valued customers, for example, in combination with a WAP-enabled device.

For a WAP site to be interesting to advertisers, it is important to collect as much information on the users as possible. The more data available on the users, the more targeted advertising can be. The more the visitors of the site can be segmented, the higher the advertising revenue will be; that is, if you do not make this information available to third parties, and only make this information available in a generic manner.

Some portals will want to pay for your information or application, especially if it offers a greater added value due to the fact that the information is up to date, the application is easy to use, or the information cannot be obtained cheaper anywhere else. For this, prices per customer can be calculated, but also charge a lump sum per month. In general, licenses per user in the case of software applications can yield much revenue.

Place

The accessibility of your site is an important consideration. The ways to generate visitors do not differ that much from those used on the Internet. The biggest differences are the limited space, for example, for banners and the troublesome entry of URLs via the keyboard of a mobile phone. If a unique customer identification, invoicing the end user, or location information are important to your service, now or in the future, it is wise to join well-known portals. Mobile operators already have a customer relationship with the visitors of your site and are able to provide customer identification and invoicing to the customer as part of their portal service. Parties such as Internet portals and providers of free Internet do not have such relationships. Mobile operators could supply additional information. The most valuable appears to be location determination.

A mobile operator has more relevant information at its disposal. The value of real-time device identification, browser identification, and network type identification gives these operators an opportunity to supply the visitor with information tailored to the situation. Device type is important for the size of the screen and the screen resolution, and whether or not the screen displays color. Browser identification is also an important factor. With the development of WAP, new browser versions will be launched with more possibilities. At the time of publication of this book, it was not yet possible to download a new browser version via the mobile telephone. This means that there will be several browsers used, comparable to several versions of Netscape Navigator and Microsoft Internet Explorer. Network type identification is also important as users will approach your service via "regular" GSM, GPRS, or 3G. The different carrier services offer different data speeds, comparable to analog, ISDN, and ADSL on the regular telephony network. Your service will have to use the possibilities of the device. Watching black-and-white television on a brand new, big-screen, color television is not appreciated by viewers. It remains to be seen if foreign operators will pass on the location of roamers on their network for free. If they will, it remains to be seen if all will do so in a uniform manner. A mobile operator will be better equipped to offer a uniform international location determination. This is important, as it will allow you to either use another portal or have a WAP site that can be reached with one URL. To draw visitors to your WAP site, it is wise to be connected to a portal that advertises a lot. Compare it to a department store or mall: You will be better off if your store is located in a busy mall than on some side street where nobody ever goes.

If your site can be reached via several portals (for example, because you offer a free service), it is important to claim a good URL. An option is to use mmm or wap instead of www in your URL. For this you do not have to register for a new domain name. Bruna, a Dutch bookstore, for example, uses *mmm.bruna.nl* for WAP and *www.bruna.nl* on the Internet. Entering a URL is not always easy for a user. Some suppliers of mobile Internet will also block entering a URL (see Chapter 5) so that users only have access to the services they offer.

Registering with search engines is also wise. Select good keywords that describe the nature of your site. There are also many different portals and start pages on the Internet that review interesting WAP sites (see Table 9.2).

Table 9.2
Common Search Engines and Portals on the Internet

Search Engines and Portals on the Internet	
www.wapaw.com	*www.yourwap.com*
www.boltblue.com	*www.wannawap.com*
www.wapmap.com	*www.ccwap.com*
www.nl.wapjag.com	*www.mywaplink.com*

It is important that you can be found where the user expects you to be. In a news portal, the online edition of the daily and weekly papers can be found. A search portal (see Figure 9.2) such as Raging Search also offers search services for specific products or services. Think of a search engine that will help you locate a cheap laundromat in your city. The same goes for order and reservation services. For search, order,

Figure 9.2
Raging Search, a search engine for wireless content.

and reservation services, specific portals and search engines are very important. By specific we mean portals aimed at young parents, do-it-yourself people, stamp collectors, and so on. In the business-to-business market, we think of branches, chains of buyers and salespeople, or a collection of products and services in a city or state. Users searching for entertainment will land on sites by familiar icons such as Disney, Playstation, and Nintendo. New brands of the mobile games segment will appear, such as the Japanese Bandai.

Promotion

A well-known communication model is AIDAS: attention, interest, desire, action, and satisfaction. First, the potential customer has to be made aware of the product you offer (Figure 9.3). Next, the potential customer has to become interested in the product and their desire to purchase the product needs to be reinforced. Then the customer needs to be encouraged to take action and purchase the product. After the purchase, the user will determine whether or not he or she is satisfied with the product and what the likelihood is of purchasing the product again in the future. For products you buy in stores that are promoted via traditional media, such as newspapers, television, radio, and magazines, the time and place of promotion does not coincide with the time and the place of purchase.

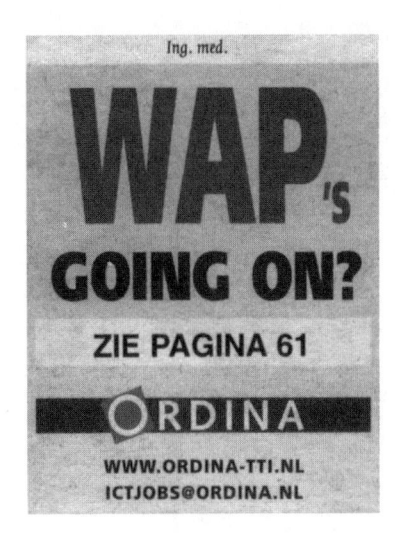

Figure 9.3
Example of a hiring campaign by Ordina.

With in-store promotions and promotions on packaging, this is the case. The time and the place do not always coincide with the time and the place of the consumption or the use of the product. With WAP, time and place of promotion can coincide with time and place of purchase and even with time and place of delivery and consumption; for example, soda machines and parking meters. This means that the whole cycle of attention to purchase and even use is shorter. Let's say you are waiting for a train and you see a poster with an announcement for a show you would like to see. Your interest is triggered. The poster reads: "Reservations: mmm.broadway.com." You can make your reservations directly via the mobile phone, and you can access your schedule to check if you have any prior engagements. You can make your reservations, pay, and maybe even have the tickets sent to your mobile phone or pick them up at the venue. In less than five minutes, the attention of the customer was triggered and the purchase was made. If the venue is close by and it concerns a last-minute offer, reservations could be made shortly before the show starts.

The promise that is made in the communication, picked up by the customer, is rapidly tested by the customer. The promise has to be fulfilled. This shortened communication cycle and the possibility that time and place of promotion coincide with the purchase have consequences for your choice of media. It is very important to use the media, which allows the customers to take action right away. A referral to your current news site in magazines or newspapers is effective. Small and relatively cheap editorials lead to a loyal following of, for example, sports-interested users. An advertisment of your search service in paper media such as the Yellow Pages or the tele-

Step 2: The Marketing Mix   237

phone directory relates to the experience of the target audience, similar to the way advertising for your order and reservation service relate to your own magazine or trade journals. An advertisement for a mobile game in the *Metro*, a free newspaper distributed at train stations, could be very effective, as readers using public transportation are often bored.

Outdoor advertising is especially intriguing for WAP services. If users waiting for the bus are triggered by your advertisement, chances are they will visit the site right away. The same goes for outdoor advertising at special events, such as concerts or amusement parks, where waiting times are long. Advertisements in newspapers and magazines can also entice readers to take action. As opposed to the regular Internet, your WAP site can be accessed immediately. The chance of direct response partly compensates for the inconvenience of having to type in the URL on a small screen. The success of SMS shows that this is not a problem for the younger generations, regardless of the price.

Internet as a promotional medium is important, too, first because of the overlap in the target market between Internet and WAP. Second, the multiaccess aspect of your service, both via Internet and WAP, can be emphasized. Third, customers can be informed via the Internet, so that use of the WAP service becomes easier and faster. The downside of the Internet is that in the first few years customers will not have a reason for taking action with regard to WAP. In the long run, mobile Internet could fulfill the payment function for the Internet. This means that a customer can search, select, and order via the regular Internet. Payment will take place via mobile phone. WAP is therefore a substitute for secure payment software or the SmartCard readers linked to the PC. Mobile Internet will partially overlap the regular Internet and also reach target markets that have never used the regular Internet in the past, as a result of the increased speed of mobile Internet and the penetration of mobile appliances with larger screens and easier data entry.

Radio and television can also be useful media, if the customer can respond directly via WAP. Think of all the possibilities to voice your opinions. In your campaigns on television, on radio, in newspapers, and in magazines, you can draw attention to your service via WAP, for example, by printing the URL. It is worthwhile to research if a snowball effect can be achieved by offering the users of your service the opportunity to make others aware of your service via SMS or email. On the Internet, several e-zines offer the option to forward news messages to friends. For many WAP services, such as job openings, news, and directions, this could be a useful expansion of the service (see Figure 9.4).

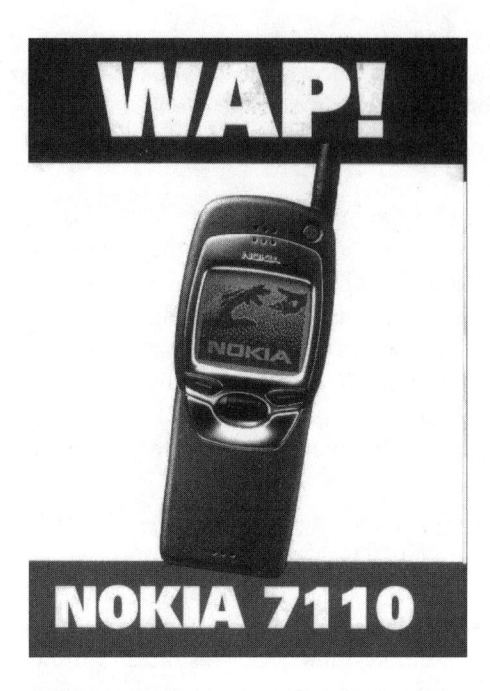

Figure 9.4
Direct mail leaflet from *Staff Planning* to attract new personnel, the golden tip earns you a Nokia 7110.

Personnel

With WAP and Internet, much of the communication with your customers will be automated. This does not mean that you do not need personnel to manage customer relations. If your customer has a problem with a payment or shipment, he or she would probably like to speak to a person. As mentioned previously, the mobile Internet user expects that the service via the telephone and the mobile Internet site are integrated, so he or she can use the same mobile phone to call and surf! Certain things are expected from your personnel. You can greet your customer on a first-name basis when your customer enters the site. Next, the customer enters data. When the customer wants to speak to somebody, he or she expects that the person to whom he or she is speaking is aware of the situation. Having to state a name again will cause irritation. This can be compared to the help desk that transfers your call after you just gave them a lengthy explanation of the problem, only to have to tell the same story again. This has far-reaching consequences for your IT environment and for your employees. Your employees will have to be able to lead the conversation to find out what may have gone wrong. Customers who are used to around-the-clock service will want to get a person on the phone the moment they visit your site. Outsourcing this process to a call-center service could help in this case. To give employees an image of what the customer experiences, employers can give their employees a WAP-enabled device.

Critical Success Factors for the Marketing Mix

To become successful with mobile Internet, the following factors are important:

- Content and portal expertise
- Distribution strengths
- Partnering ability
- Mobile technology knowledge
- Speed of innovation
- Location information
- Personalization
- Simplicity for the user
- Power of the brand
- Billing relationship toward users
- Power of communications
- Knowledge of national markets and customers
- Trust
- Flexible solutions for separation of business and personal use

Experience with the Internet (content and portal expertise) helps in setting up a mobile Internet site. Distribution is not only important for WAP, but also for other available media. Ability for partnering is a condition in an environment that changes rapidly. Keeping track of everything yourself is almost impossible. Many companies hire consultants or outsource their mobile Internet site because of the specific mobile technology knowledge needed. By innovation speed, we mean the speed at which the portal can introduce new possibilities, such as personalization, location-based services, and conducting payments. It is very important to build a loyal customer base. On one hand, you can yield higher revenue from advertising and information subscriptions. On the other hand, a strong negotiation position with merchants arises. The power of the brand has to draw in visitors, whether or not supported by communication. To have users visit your site as often as possible, not familiarity with the brand but loyalty to the brand will be the deciding factor. Trust you have already gained or still have to gain by providing good services. Knowledge of the national or even the local market is much more important for the mobile Internet than for the regular Internet. Customer knowledge will make a difference in one-to-one marketing via the mobile phone. Flexible solutions for a mix of personal and business uses will also make a difference. The employer will most likely not want to pay for personal use of the mobile phone. Services will have to be able to address these issues.

STEP 3: FINANCIAL ANALYSIS

Costs

We cannot predict the costs for your WAP site, as this depends on many factors. We can however, give you information on a few cost components and discuss your options and the extent of the costs involved.

Devices

Without a device, there is no service. The target audience for your service will have to have a mobile Internet-enabled device. It could be that you expect the target market will purchase a WAP device and that you will not have to assist in this process. It could be that you want to stimulate the purchase of WAP devices with your target audience. You can try to cooperate with a device manufacturer or a mobile operator to reach the target audience. It will be beneficial for a device manufacturer or mobile operator if you can reach a large target audience via direct channels such as direct mail, and you take on part of the expenses incurred for similar actions.

If you already have a savings plan with your customers, such as air miles or something similar, you can have your customers save for a mobile phone. If you want to equip employees with a WAP telephone, it is advisable to use the moment of contract extension with your mobile operator if you already have mobile connections. You can often purchase devices at lower prices than if you were to purchase the devices without subscriptions. Check if the new devices suit the accessories on hand, such as car kits and data cables. These are often hidden costs that can increase rapidly.

Gateways

If you decide to offer your own WAP service, you can opt to purchase a gateway. Using your own gateway means you can determine the rates for traffic at which the customer dials in. You can choose to use a toll-free 1-800 number, a paid 1-900 number, or a regular number. You also have more influence on the security of the gateway and

Step 3: Financial Analysis   **241**

the connected systems. This could be a temporary advantage, as the mobile operators and ISPs work hard at the security of their networks, their gateway, and connections between their gateway and your systems. The question is whether these advantages will outweigh the disadvantages. Users will have to program your gateway's settings in their devices to use your service. Most devices only offer space for five different gateway settings. If your service is for your own employees, this will be different. You can facilitate your employees more easily in setting up the device (often the same types). For customers with many different devices, this is more complex. For customers, you will probably incur increased communication cost if you have your own gateway compared with making your site accessible via portals.

The cost of the purchase of a gateway and its maintenance are much higher than offering your WAP service to different portals. A gateway costs anywhere from nothing to $20,000, depending on the capacity. To this you will have to add maintenance costs. If you purchase your own gateway, you will have to make sure that your gateway stays up to date with all the developments in the WAP standards and mobile networks. These costs are difficult to predict, but will be considerable due to the expected additions to WAP. Another disadvantage is the time to market. Installing and testing your own gateway takes extra time, on top of the development of the WAP service itself.

Software

You can outsource the development of your WAP service, but you can also do it yourself. The large advantage of outsourcing is that it often takes a shorter period of time for the experts to manage the project than it would if you have to educate your own personnel. However, if you already employ HTML experts and system experts, it is advisable to develop your WAP service in-house, especially if you expect regular updates to the site. For graphics design and development of the user interface, you could always seek the advice of experts.

The cost of developing software depends on the systems already used. If only a front end, a WAP user interface has to be developed and all other systems already exist and have well-written interfaces, the development will not take much time. We are familiar with examples from companies that developed a beautiful site within two weeks at a cost of 80 hours of programming at $100 per hour, for a total of $8,000. If the back-end systems, such as inventory management and customer administration, will have to be adapted, the cost could be much higher. It is advisable to involve experts in the early stages of development so that they can estimate the impact the new WAP service will have on the existing processes and systems.

Some Internet sites offer the opportunity to create a WAP site on the spot. You can already create your own home page at *www.wapdrive.com*. There are WAP site tools available aimed specifically at companies. The costs involved are virtually none or very small. Often the supplier also hosts your site and provides you with a URL. For simple sites with information that does not require links to your systems, this could be an attractive and inexpensive option.

Hardware and Connections

You can run your WAP service on your own hardware, but you can also have it hosted by a Web hosting company. The costs of hosting are usually minimal, about $10 per month. If you already have your own Internet site running on your own hardware, adding your WAP service will not increase your cost, as the bandwidth and memory needed are much smaller than that for an Internet site. An important parameter in the decision of self-hosting or outsourcing is management. If you offer WAP and Internet services, you have to make sure that your systems are up and running at all times. If you are not able to provide this, outsourcing WAP hosting is the solution.

Another important hardware aspect could be the installation of GSM modules in appliances. If you want your customers to pay for a soda via a WAP service, payment will have to take place directly on delivery. This can be done by connecting a processor and a GSM module to a previously installed processor in the appliance. Installation of such appliances means that stock and failures can be controlled remotely. Per appliance, you will have to count on a cost of $1,000.

Marketing and Promotion Expenses

Your marketing and promotion expenses depend on the means at your disposal and the frequency of the activities. In general, new and innovative services receive much attention from the press and the initial promotion expenses could be low. You can draw the attention of existing customers to your WAP service via regular means of communication, such as direct mail, in-store promotions, or newsletters.

Fulfillment

Costs of fulfillment consist of those costs incurred to supply the product or services. To set up the fulfillment process so that the product can be delivered on order, as in the case with vending machines and ticketing (Broadway tickets, airline tickets, etc.), is often a costly procedure. A GSM module will have to be installed in the vending machine, and new software will have to be written and hardware will have to be purchased to manage delivery. For ticketing, a system will have to be developed with which the customer can obtain tickets or gain access to the show using a password or code. If you offer last-minute reservations, customers do not want to wait in line to get the tickets.

Shipping and handling charges will also have to be taken into account. Often, customers have to pay for shipping and handling separately, which hinders the purchasing process. It is often easier to calculate these charges into the price of the product so that the customer can make a better comparison between buying the product via WAP or buying the product in the store.

Other important costs are the costs associated with payment. There are many different transaction providers that can facilitate payments via WAP. Often, a fee is charged for each transaction. For existing customers you already send invoices to, you can charge their WAP services and orders via these invoices, so the additional costs will be small.

Education

In some cases, employees or customers will have to be educated in the use of the service. This can be done at low cost, if an educational demonstration is developed that can be used via the Internet or via a CD-ROM or floppy disk. In this case, employees or customers could use the education demonstration, or use it as a reference, when it is convenient for them. For smaller companies, this is not a sound method, but in this case the WAP activities will be limited to a smaller number of people. With regard to technology, much information can be found in newsgroups and on sites such as *www.wapforum.org* and *www.nokia.com*. Similar to the Internet, a number of consultants, education centers, and communications agencies will add mobile Internet to their portfolio. Finally, numerous conferences are being organized by telecommunications companies, consultants, or independent congress organization bureaus specifically in the area of mobile Internet.

Revenue

Besides the costs, the expected revenue will also have to be mapped out. This can be very simple: Consulting work schedules via WAP will save the employee about 20 minutes per day in commuting time. If the employee can use this time in another way, for example, by serving more customers, this time will indirectly increase revenue. In other cases, this is more difficult and other assumptions will have to be made. We use the example of buying and selling stocks via the mobile phone. Using general market numbers, it can be estimated how many people have a WAP phone. Next, we will have to determine the percentage of people investing and the percentage of people that use their bank for buying stocks. Next, we have to estimate the number of transactions that will be made via the mobile telephone and the percentage of extra transactions, as well as regular transactions the customers would make regardless. With the average size of the transaction and the percentage the bank charges as commission, the revenue can be estimated. This includes many assumptions, in which small inaccuracies could lead to very different outcomes.

The revenue can be divided into a number of categories:

- The user pays for your information
- The WAP advertising aimed at your specific user profile brings in revenue
- Referral fees for generating revenue for other sites that are reached via your site
- Additional traffic from existing customers
- Additional traffic from existing customers because the users see and buy other products using WAP instead of ordering them over the phone
- Faster payments, as paying time via the mobile phone can be reduced from 30 days to 2 days
- Cost savings

A WAP service can save you money. A WAP service for customers could function as a customer self-care instrument, decreasing the number of questions via the telephone. You might even receive more questions that lead to sales.

Saving on negotiations between salespeople and back-office personnel, faster offers and service traces toward customers will likely yield revenue.

The question is how much the revenue really is. It is therefore advisable to test new services. To test a service, often only the front end of your service has to exist. An employee could take care of the back end of the service during the test. Testing with real customers often gives a realistic image of the possible revenue of the service. At this time, we can also calculate whether or not it would be worthwhile to equip your customers or your personnel with a WAP-enabled device. Some effects, such as image improvement and a better reputation, are more difficult to express in dollar amounts. The same goes for customer satisfaction of your most important customers, so that they remain your customers. However, it is smart to label these effects so that it is measurable to which extent they take place.

Calculations for possible mass-market applications rise or fall with the penetration of WAP telephones and user acceptance of this new medium. Maybe the numbers mentioned in this book will aid you. Also, a comparison of startup time and use of similar services for other media such as teletext, Internet, and television provide useful information. The downside of entrepreneurship is that the moment it becomes clear a service will make money, somebody else is already active in that space.

STEP 4: PLAN OF ACTION

After you formulate an objective, fill in your marketing mix, and perform the financial profit-and-loss analysis, you can determine your plan of action. If you decide to develop a WAP service, you will have to define what you want to offer at introduction and what you can offer in releases to come. Of course, the minimum service has to have mobile relevance. Publicizing a telephone number via WAP is done in no time, but does not offer users what they expect of your WAP service, at a minimum. It may be better to provide the user with a selection from a limited assortment of products he or she can order right away that will be delivered correctly, as opposed to providing information on a complete product line without the option of ordering any of these products.

We advise you to work with releases for the following reasons:

- *Fast time to market.* To realize all product wishes, more time is needed as opposed to offering the minimum basic service.
- *Adaptation.* The moment you offer a basic service to the market, you can test this service among users and collect feedback on your service. You can use these suggestions in further development of your product.
- *Communication moments.* With every release you will have something new to report: an even larger product line, more features, and faster deliveries. This will give you the opportunity to provide the press and your customers with new information.
- *New technological possibilities.* Working with releases will most likely be needed, as new technological possibilities that can be used for further product development arise. By working with a release, you can slot in time for taking advantage of the technological advantages.

STEP 5: CONTROL AND REPORTING

An important part of formulating the objective and the financial analysis is formulating benchmarks against which the success of the service can be measured. These could be financial benchmarks, such as turnover and profit expectations, but also communication objectives, such as familiarity with the service, brand image, and so on. Other customer satisfaction objectives can also be formulated, such as, "More than 90 percent of the users of our service are satisfied or very satisfied with our service." It is very important that the objectives are specific and measurable. In the case of Internet technology, much can be measured, such as the number of users that have used the site, the number of pages consulted, and even the browser used. The number of successful transactions, the number of ads displayed, or the number of referrals to other sites can become important elements for direction, dependent on the objective. Of course it is advisable to ask for regular feedback from customers so that the site can better serve them.

10 A Look into the Mobile Future

Many books on new opportunities, technologies, or methodologies start with a beautiful vision and then cover what can be done now and in the future. This is often disappointing. We have tried to indicate what is already available in this field, and which concepts you as an entrepreneur can use to gain a competitive advantage and take advantage of the trends in society. However, we do want to give you our vision and thoughts on the developments in information, communication, entertainment, and transactions for the future. It could very well be that the future will be completely different from our expectations. So be it. This is not about right or wrong; it is about us getting you to think about mobile Internet. We plan on a future with several services and applications to make life more enjoyable, a future in which you can make money with the company you work for now or with the company you might start in the future. Numerous descriptions have been presented of the houses and offices of the future. You will not be surprised to see that in the scenario described here, mobility plays a central part instead of a house or an office. For that reason, we selected "a day in the life."

Work starts at home in the morning where I check my email and day planner after the morning paper at breakfast. I have an appointment with a customer in the middle of the city. I decide to take the subway and for the last part of the trip catch a ride to the office with a coworker. I send her an email and receive a reply that this is not a problem. She was stuck in traffic. I switch on my mobile phone (Figure 10.1). I don't even

Figure 10.1
Nokia prototype for a terminal with video conference capabilities.

have to use a PIN code anymore, but instead activate the mobile phone by pressing my thumb to the side of the phone. This way I can also write my electronic signature.

I reserve a parking space using the mobile phone, as I already switched off my computer. Payment takes place automatically, so I do not have to worry about getting a ticket in case the meeting runs late. I have to hurry. After parking my car in the reserved spot I get on the subway without having to wait in line for a ticket. Even though I do not travel by subway often, I have a subscription that allows me to only pay for actual use of the subway. Thanks to the Bluetooth chip in my mobile phone, the transportation authority registers when and where I board and where I get off. Every month the cost of my public transport journeys are calculated and taken out of my bank account. I can also pay for my cup of coffee using my mobile phone, which is easy when I run out of change. Once a month I log on to my bank's site and indicate which expenses were business related, so that I can easily do my expense reports. At the station, I also drop off my dry cleaning. In the subway, I read the new comic strip I receive every day on my mobile phone, and I use my laptop to prepare a presentation I have to give in London tomorrow. Luckily these days the mobile coverage in subways is good, so there is no problem sending and receiving data via the mobile phone. My presentation deals with new services for the travel industry. I miss a few funny quotes and remember an interesting interview on television with one of the hot-shots in the travel industry on the importance of service. Fortunately, there are numerous good sites with clips from TV shows and movies. I give my search agent the assignment to find the interview. My search agent finds not only that interview, but also another interview with a different person in the travel industry. I decide to use both clips in my presentation. I send the presentation to London so that they can have everything ready for tomorrow.

I have to find out from a coworker, who also is flying to London tomorrow, what flight he is on. I do not know his number, so I speak his name into the telephone. There are more coworkers with that name, so I receive pictures of the people from the intranet. I search for the right person and call him. He is in a meeting and receives my message on the screen of his mobile phone. He writes back that he is on the 8:00 a.m. flight. I am on a later flight. I check the airline to see if I can be on the earlier flight. Fortunately, there are seats available, so I reschedule my flight and receive a confirmation.

I also receive a message that my secretary is celebrating her birthday today, so I order some flowers (she normally does this for the other coworkers). I send out a group email right away to our department to have everybody send her a Happy Birthday e-card. While I get out of the subway, I receive a call from the coworker who's picking me up. She calls to ask for directions. I am not too familiar with the area, but give her directions based on the mobile "find your way" service. Together we go to a meeting with a client.

During the meeting a number of new issues are raised. To keep the momentum going I search for the status of the project on our intranet, which may provide a solution to the problem. As the site does not answer all my questions, I call the project manager via the internal phone list, and ask him for clarification. He is introduced to our client and clarifies the remaining issues via a video conference. At the end of the session I schedule a follow-up appointment with the project manager and the client using my organizer connected to our intranet, which allows me to check the schedule of the project manager. While I do this, I receive a message that my second appointment of the day has been canceled. I accept the proposal for a new appointment. It then becomes clear that my secretary has scheduled another appointment for the same time. I pick another date right away. Back in the car I send my notes from the meeting from my laptop to the project manager, so that he is aware of the situation and can start working on the action points.

My coworker and I decide to have lunch in the city. She has heard about this new café. Via my mobile phone I find the address and get directions. Walking is faster, according to the mobile agent. As it starts to rain we decide to take a taxi. Taxis are never there when you need them. Using my mobile phone, I find the number for a local taxi company and order a taxi. I pay for the ride using my mobile phone. At the café I see a familiar face, but I cannot remember his name. Via my contact list with pictures I find the name of the person. Lunch is great and I forward the address to my secretary. She keeps a list of good restaurants, in case we need to take a client out to lunch. She emails me to let me know that she will have a little birthday celebration in the office in one hour. In the meantime, it has stopped raining and we walk back to the car.

On the road I realize that I have to take my own car to the garage during my next business trip. On my garage's Web site, I find that this is indeed the case and I make an appointment right away. I receive a reply that it will soon be time for my annual safety test. As it is cheaper to have this test done right away, the garage asks me to bring the paperwork. I reply I would like them to pick up the car as well and suggest a time. They confirm it immediately. My coworker reminds me of a great show in London. If you have to be there for work you might as well have fun! I mail my travel companion a message, asking him if he wants to come along. Via the mobile phone, I search for the show and reserve two tickets. If my coworker has other plans, a cancellation can be done right away. The reservation line sends me some tips for local restaurants. I decide to store them for a while; you never know. Back at the office we are just in time for birthday cake. The project manager has already started working on the issues I discussed during the meeting this morning. He wants to know more about the meeting so that the proposal can be mailed out today. When I sit behind my desk and boot up my laptop, I check if my coworker in Japan is available. I see that he can be reached via mobile phone.

A Look into the Mobile Future

Figure 10.2
Sony prototype for a terminal with photo camera capabilities.

I first type a message to determine if I can contact him. He sees the message right away on his mobile phone and he replies that I can call him. We talk about his stay in Japan and his progress with our Japanese customers. He saw a really great new gadget, a small camera that you can connect to your mobile phone (see Figure 10.2). You can record short movies and send them via the mobile phone. He also saw a gadget that can record and transmit scents. A PC or mobile telephone user with a scent module can smell scents recorded elsewhere. We decide to hang up and brainstorm the opportunities for such a product for other markets outside Japan while chatting. (He talks into his mobile phone, which is transformed into text for me behind my computer. After 10 minutes we are done and have compiled a list of almost 50 ideas.

A colleague wanders in with a competitor's press release that confirms a rumor that the competitor has found a foreign partner. Together, we check out the partner's Web site. I see on the screen of my mobile phone that the rumors have put pressure on the stock prices. I ask a colleague to put together a quick analysis of the possible results of this cooperation and the steps we have to undertake so we can discuss this in the upcoming weekly management team meeting. I decide to take the mail home and to leave now to stay ahead of rush hour. On the subway, I read most of the mail and send some email. I call my spouse to say that I will take care of dinner. I order the missing ingredients directly from the grocery store. I receive a confirmation of the order and a pickup number right away via the mobile phone.

At the metro station I pick up my dry cleaning, pay using my mobile phone, and drive to the supermarket. My bags are ready. By showing my pickup number, I am certain to take away the right groceries. As I pass a mall, my favorite leisure shop sends me a message that their new collection has arrived. Even though I would like to check it out, I decide to drive home. At home I scan the proposal and talk to my colleague on the phone before I mail the proposal to the customer. The mail brings me a number of bills and while I cook supper I pay the bills via the mobile phone.

The food tastes good and I watch the news afterward. I hear a beep from the mobile phone and see that my parents mailed me a nice vacation picture from the South of France (see Figure 10.3). They seem to be having a great time! After the news, an interesting sports game comes on. Viewers are encouraged to place a bet via voice response or via the mobile phone. I bet on a score. After the match (wrong score, bad luck) I pack my suitcase for tomorrow and chat with my spouse. Before I go to bed I quickly order a taxi to take me to the airport in the morning. My travel companion has sent me a message letting me know where to meet at the airport and to let me know he would love to join me for the show tomorrow evening in London. Great!

Figure 10.3
Example of a postcard (Nokia prototype).

Appendix A: Examples and Tips to Build a WAP Site

A simple way to create a Web site is to visit a WAP home-page construction site on the Internet. Such a site offers a simple user interface for building a WAP site. They often also offer graphics and show how the WAP site will look when accessed via a mobile phone. On the following Web sites you can find WAP home-page builders:

www.wapsilon.com

www.celon.net

www.wappy.to

These sites are very suitable for private and new WAP sites.

For a professional WAP site, it is often necessary to have a WAP site programmed. It is useful to download a WAP toolkit from the Internet. There are many different sites available, often linked to a specific type of WAP browser and WAP device. You can download WAP toolkits from the following sites:

www.forum.nokia.com

www.ericsson.com/WAP/developer/

These toolkits also often include a WML manual, which explains how you can program for WAP. In the figures in this appendix, we show some examples of WAP sites and related WML code.

WML code consists of decks with cards. The moment a user selects a URL from a WAP site, the deck of cards is sent to the mobile devices at once. Within a deck, cards can be clicked through rapidly. As soon as another deck of cards or a different URL is selected, the mobile device needs time to download the next deck of cards off the network. There is a limit to the size of a deck, but that depends on the browser and the device. Table 1 is a list of the most common browser and device types.

Table 1
Most Common Browser and Device Types

Browser	Limit to Compiled WML Deck
UP.Browser 3.2	1,492 bytes
UP.Browser 4.x	2,048 bytes
Ericsson R320	About 3,000 bytes
Ericsson R380	About 3,500 bytes
Ericsson MC218	More than 8,000 bytes
Nokia 7110	1,397 bytes

The UP Browser is used in Motorola devices.

Here are some tips for the user interface of a WAP site:

- Keep it simple! The user interface options of a WAP application are limited, the screens are small, and the design possibilities are very limited.
- Keep text short and simple. Reading from the screen of a WAP phone is difficult, mainly because scrolling does not give an indication of the length of the text and the location of the cursor.
- Offer an alternative for the standard "accept" menu option via the "options" menu for a WAP application for inexperienced users by using a hyperlink in the text.
- Prevent the user from having to enter too much text. Let the user choose from a list of options where possible.
- Limit the use of pictures, unless it adds value. A logo when entering the site is acceptable if it offers the option to continue.

You can find programs on the Internet that can convert bitmaps into wireless bitmaps. Check, for example these sites:

www.gingco.de/wap/

www.teraflops.com/wbmp/

A plug-in also exists for PhotoShop and Paintshop Pro with the name *Unwired*:

www.rcp.co.uk/distributed/Downloads

In using pictures, it is important to know that not every phone has square pixels. The screen of a Nokia 7110 does not have square pixels. As a result, the pictures on the screen of a PC are perfectly round, whereas they are oval on the screen of the

Nokia. In designing pictures using a PC and adapting the screen size, pictures can be created that can be seen on a Nokia 7110 screen without distortion.

The following examples give an indication of the possibilities of WAP 1.1. The screen captures display a WML code for the screen on the telephone. These examples have been provided by iMedia.

EXAMPLE 1: A FLASH SCREEN WITH LOGO..............

In this example, the iMedia logo is shown for five seconds (50 * one tenth of a second) on the screen of the WAP device (Figure 1). The logo is a wireless bitmap named i.wbmp. When the timer runs out, the user will be connected to the URL: *mmm.imedia.nl/start.wml*. A screen that disappears automatically after a while is called a flash screen.

Figure 1
Showing a logo for five seconds.

EXAMPLE 2: LINKS TO OTHER PAGES

This example shows how you can incorporate links to other pages (Figure 2). This link could be to a card in the same deck or to a different URL. Every card can be named, which can be displayed on the top of the screen. If the user clicks on the icon link to card, card 2 is displayed. If the user opts for link to URL, he or she will be directed to the URL *option2.wml*.

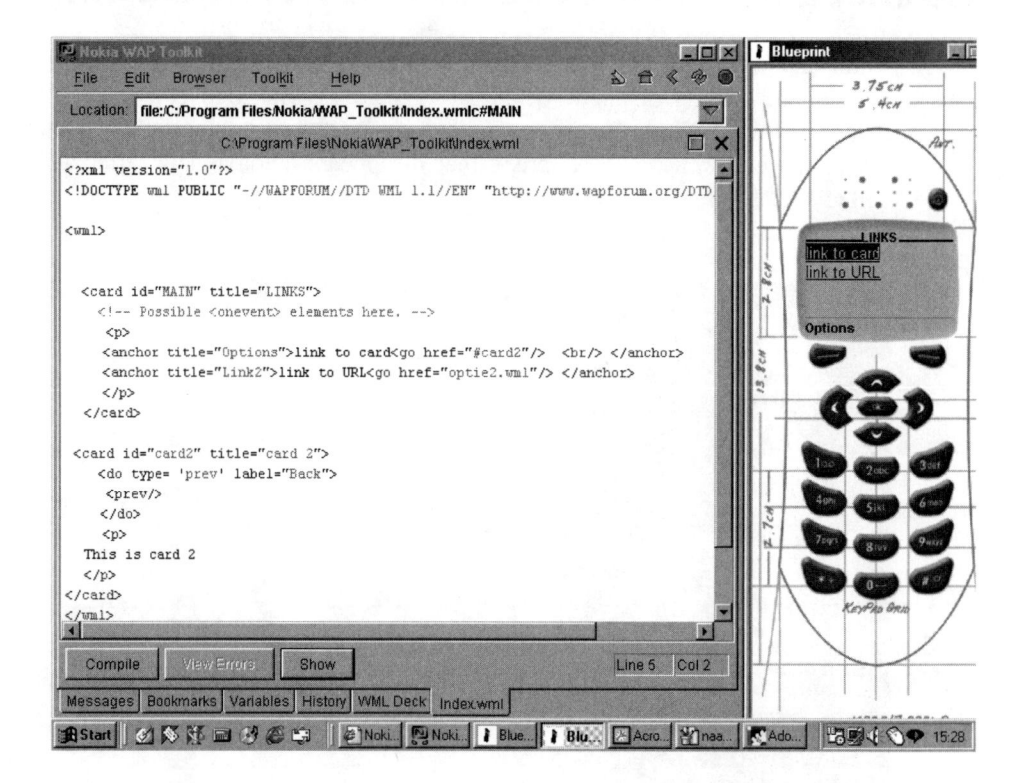

Figure 2
Showing a hyperlink on a page.

EXAMPLE 3: THE USE OF A PARAMETER

In this example, a name is entered. After the text "enter your name," a field will be displayed [....]. If the user clicks on this field, it can be edited. By format= "*M", we mean that the user can enter an unlimited number of characters. It is also possible to limit the characters to numbers or letters only. This text is then displayed in the field, for example [pieter]. The text in the options menu can be programmed so that the user can indicate when he or she has finished entering data. The example uses three screens (Figures 3a, 3b, and 3c).

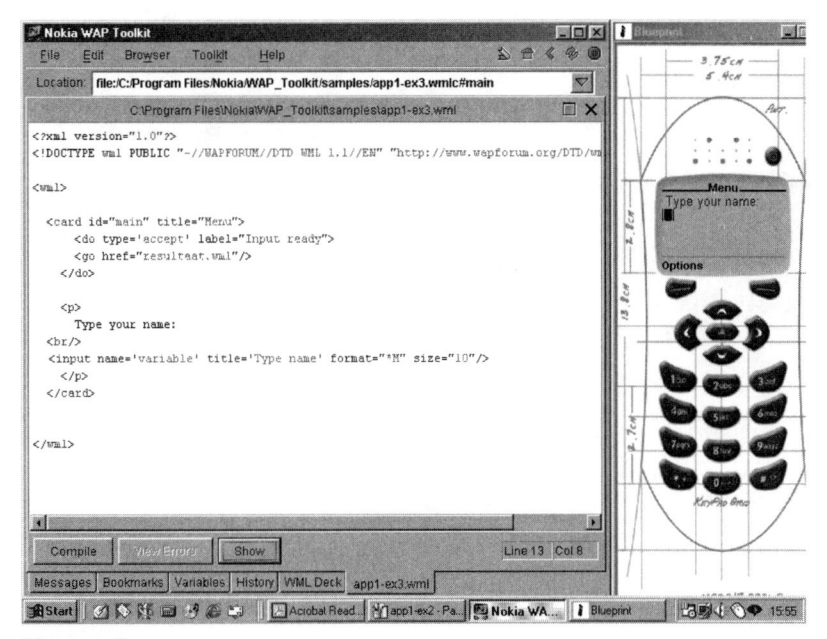

Figure 3a
An input field.

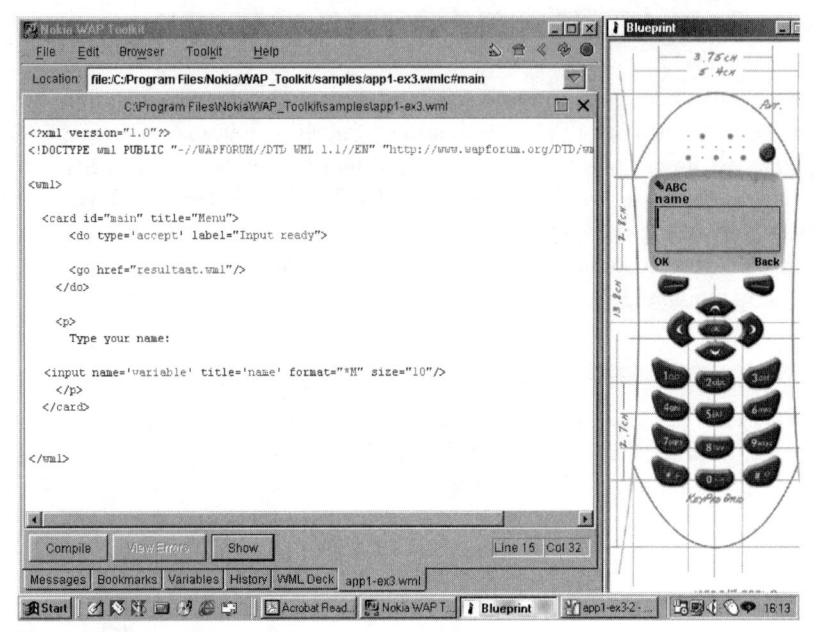

Figure 3b
Input of text by the user.

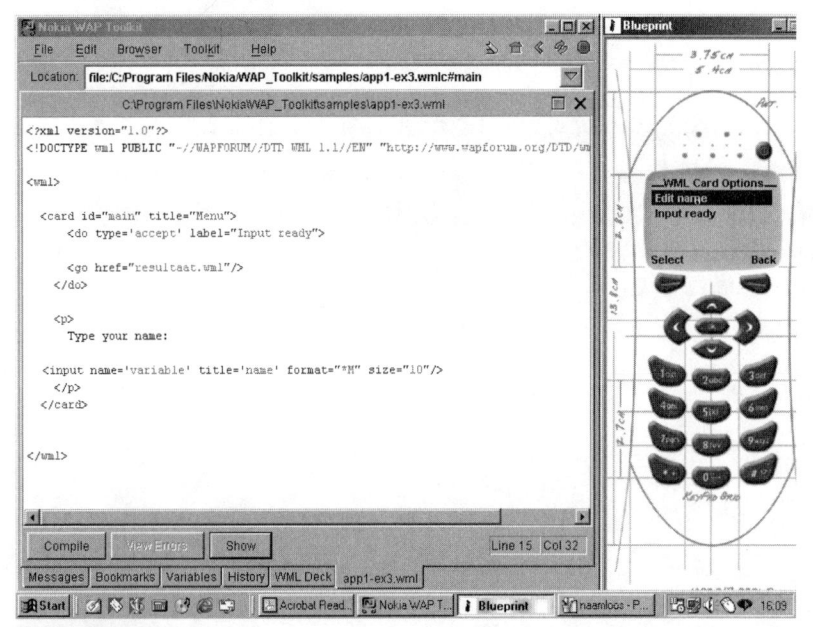

Figure 3c
Input options.

Example 4: Selection from Options

EXAMPLE 4: SELECTION FROM OPTIONS

In this case, the user, who wants to order a pizza, can select from a number of options. By clicking the field again, [...] the user will reach the options. The user can select from three different pizzas. By scrolling through the list and selecting the pizza, the user can make a selection (Figures 4a, 4b, and 4c). These lists of options allow the programmer to limit the user to selecting one or more options.

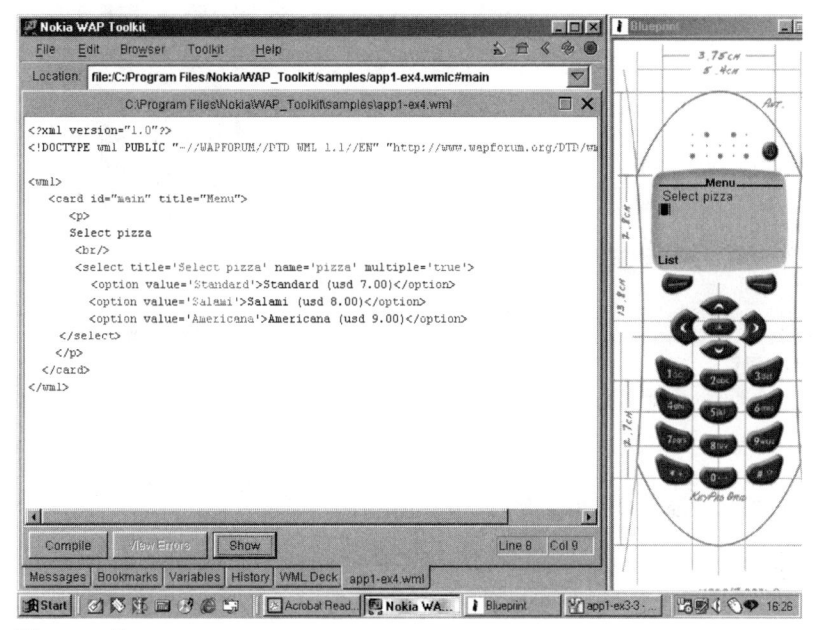

Figure 4a
A select option.

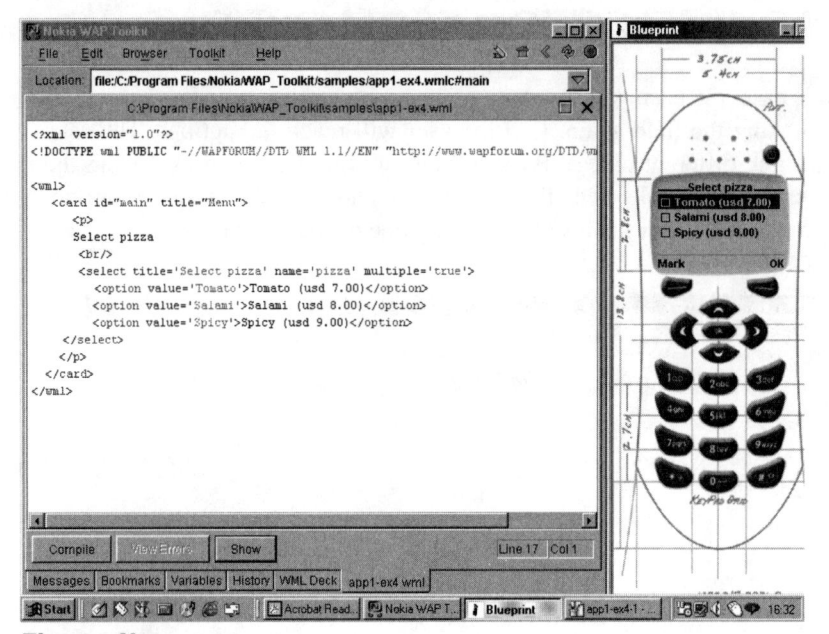

Figure 4b
Showing different pizzas.

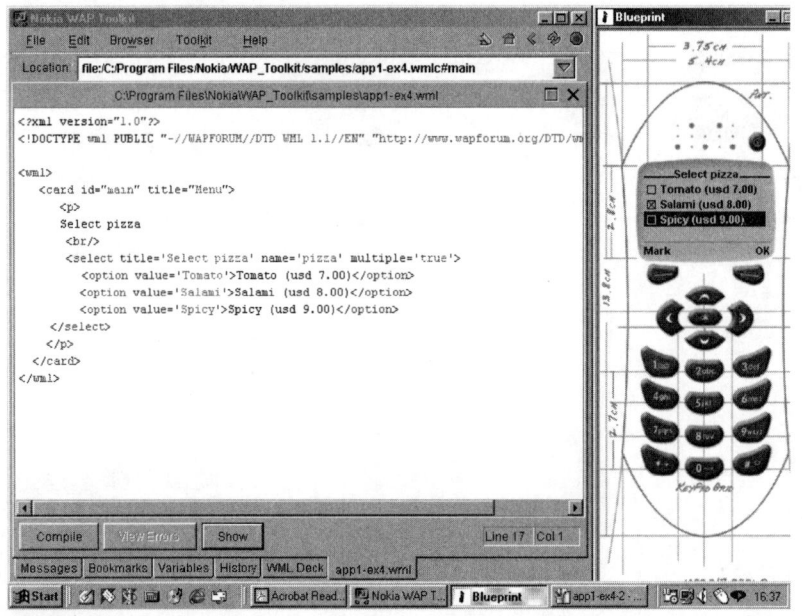

Figure 4c
Selecting a pizza.

Appendix B:
List of Interesting URLs

Information about WAP

www.allwap.com
www.waplink.nl
www.wapdrive.net
www.gelon.net
www.wap.net
www.durlacher.com
www.questus.com
www.forrester.com
www.herring.com
www.webtop.com
www.mobilestart.com
www.wapjag.com
www.wapsight.com

The WAP Forum

www.wapforum.org

Bluetooth

www.bluetooth.net

www.nokia.com

www.ericsson.com

Make Your Own WAP Home Page

www.wapdrive.net

www.waplink.nl

WAP Developers Toolkits

www.forum.nokia.com

www.ericsson.com/WAP/developer/

developer.openwave.com/

Converter for Bitmaps

www.gingco.de/wap/

WAP Suppliers

www.openwave.com

www.nokia.com

www.ericsson.com

www.mot.com

www.lucent.com

www.oraclemobile.com

www.mytimeport.com

PDAs

www.symbian.com
www.casio.com
www.psion.com
www.palm.com
www.microsoft.com
www.pocketpc.com
www.ausystem.com/wap
www.visor.com

Mobile Operators

www.nttdocomo.com
www.cellnet.co.uk
www.sonerazed.com
www.virgin.com
www.vizzavi.com

Content

www.yahoo.com
www.infospace.com
www.reuters.com
www.portalwap.com
www.yodlee.com
www.123internet.nl
www.quios.com
www.onlineagenda.nl
www.bruna.nl
www.digitalbridges.com
www.excite.co.uk
www.yourwap.com
www.boxman.com

www.wapterror.de
www.ncbdirect.com
www.bbc.co.uk/mobile

Consultants

www.capgemini.com
www.cmg.nl
www.camcom.com
www.ericsson.com/letswap
www.igic.com

Abbreviations

ADSL
Asynchronous Digital Subscriber Line, technology to realize higher transmission speeds using common telephony cables

ANSI
American National Standards Institute

B2B
Business to Business market

B2C
Business to Consumer market

Bluetooth
Radio technology designed for communication between devices at a relatively short distance, name is taken from the Viking king Blatant

CDMA
Code Division Multiple Access, radio code technology for mobile networks, mainly used in the Americas and Asia

CHTML
Compact Hypertext Markup Language, mobile programming language to create mobile Internet sites; a simplified form of Internet HTML markup language introduced by NTT DoCoMo of Japan

CRM
Customer relationship management, software and systems to store and analyze customer data to manage and optimize customer contacts

EDGE
Enhanced Data Rates for GSM Evolution, technology to realize higher transmission speeds using a GSM channel

EDI
Electronic Data Interchange, standard format for sending and receiving electronic forms

EFR
Enhanced Full Rate, technology to enhance voice quality of GSM networks

E-OTD
Enhanced-Observed Time of Arrival, method for positioning in a mobile network using triangle measurements

EPOC
Operating system for mobile terminals, like PDAs, developed by Psion

ERP
Enterprise Resource Planning, software and systems supporting management of company resources, often modular based including modules aimed at supply management, project planning, and process management

ETSI
European Telecommunications Standards Institute

FTP
File Transfer Protocol, protocol to send files over the Internet

GIF
Graphics Interchange Format, file format for images

GPRS
General Packet Radio Services, technology enabling packet-switched communication in a GSM network, using available bandwidth for data traffic in a more efficient way, offering the user higher transmission speeds and the possibility to always be online

GPS
Global Positioning System, satellite system designed for precise positioning

GSM
Global System for Mobile communications, mobile digital telephony standard (leading in Europe, Asia, and Australia)

HSCSD
High Speed Circuit-Switched Data, method to use more than one GSM channel to offer higher transmission speeds than 9.6 kbps for data applications

HTML
Hypertext Markup Language, programming language to design Web sites on the Internet

HTTP
Hypertext Transfer Protocol, protocol used for Internet connections

HTTPS
Hypertext Transfer Protocol Secure, protocol used for secure Internet connections

ISDN
Integrated Services Digital Network, digital fixed telephony network

MIDI
Musical Instrument Digital Interface, dedicated hardware protocol and file format for music

MP3
MPEG (Motion Pictures Expert Group)-1 Audio Layer 3, file format in which music is offered over the Internet

PC
Personal computer

PCMCIA
Personal Computer Memory Card International Association, standard for PC communication cards for laptops, used for modems and network cards

PDA
Personal Digital Assistant, pocket-sized electronic device containing schedules, contact databases, and other functions supporting personal time management

PDC(-P)
Personal Digital Communications, digital mobile telephony network in use in Japan; PDC-P is the packet-switched version used for offering i-mode services

Abbreviations

PSTN
Public System Telephony Network, analog fixed telephony network

RSA
Rivest, Shamir, Adelman, algorithm for public key encryption

SIM
Subscriber Identity Module, Smart Card in mobile phone, required for secure use of mobile networks like GSM

SMS
Short Message Service, value-added service of GSM networks enabling users to send and receive short messages containing up to 160 digits from and to mobile phones

STK or SAT
SIM Application ToolKit, programming environment to design and build programs for use on SIM cards; these programs can simplify usage of services, for example, the menu structure for information services and banking services

T9
Text input with 9 keys, fast way of text input on a mobile phone. Every character requires just one press on a key instead of several (e.g., to select a "c" the "2" key on your mobile phone needs to be pressed three times), T9 presents the right word combining your keystrokes

TCP/IP
Transmission Control Protocol/Internet Protocol, underlying protocol (e.g., HTTP), used for virtually every Internet application

TDMA
Time Division Multiple Access, radio technology for mobile networks, mainly used in the Americas

TDOA
Time Difference of Arrival, positioning method in mobile networks using triangle measurements

TLS
Transport Layer Security, protocol securing the connection between Web servers

TOA
Time of Arrival, positioning method in mobile networks using triangle measurements

UMTS
Universal Mobile Telephony System, new generation mobile networks (also named 3G) offering higher transmission speeds for data applications

URL
Uniform Resource Locator, World Wide Web address

USSD
Unstructured Supplementary Services Data, protocol enabling short messages over a GSM network

WAP
Wireless Application Protocol, standardized protocol for client–server applications over mobile networks

WCDMA
Wideband Code Division Multiple Access, radio code technology for broadband access using mobile networks

WIM
Wireless Identity Module, application on the SIM card, assuring client security in enabling transactions

WML
Wireless Markup Language, programming language to design WAP sites

WSP
Wireless Session Protocol, binary version of HTTP, used for mobile networks

WTA(I)
Wireless Telephony Application (Interface), application or interface to trigger telephony-related functions from a WAP application (e.g., call setup)

WTLS
Wireless Transfer Layer Security, enables secure connections between handset-based WAP browsers and the WAP gateway

XML
eXtensible Markup Language, universal format for structured documents and data on the Internet

Index

Symbols

.bone 214
@Home 181

Numerics

0804 Measurements Report 37
123internet 206
1G 22
2.5G 25
2.5-Generation Mobile Networks 25
24/7 141
2G 23
3G 15, 22, 28, 114
802.11B 41
99 Lives 157

A

Acrobat 62
Additional Communication Channels 178
Advanced Mobile Phone Service 22
Advertisers 107, 140
Aibo 150
Airtouch 90
Alcatel 5, 7, 88
AltaVista 106
Amazon.com 102, 107, 108, 146, 180
America Online 17, 103
AMPS 22
Ansoff's Growth Strategies 222
AOL 17, 104, 136
Apple 41, 102, 105
Appli 119
Application layer 11
application service provider 211
Ariba 105
ASKJEEVES.com 157
ASP 211
AT&T 4, 106
AT&T Wireless 92
AtoBe 208
Atos 77

Atos Origin 199
Authentication 17, 18
Authorization 17

B

back-office 182
Bandai 119
Barclays Bank 20
Bertelsman Online 230
Bibit 110, 205
Billing 75
Blaupunkt 52
Bloomberg 119
Bluetooth 34, 39, 43, 124
Bluetooth Special Interest Group 39
Boeing 109
BOL 230
BoltBlue 136, 144
Boo.com 109
BotFighters 154
brands 88, 146, 180
brick-and-mortar 174
Bright 94
British Telecom 91
browser 253
Browser Suppliers 126
Bruna 203
BT 106
BT Mobile 20
bulk calling 87
Buyers concentration 174
Buying Power 109

C

call detail record 75
Call me Now! 54
Canal+ 133
Cap Gemini 77
car kit 95
card 12, 253
Carrefour 109
Cashing Out 164
Casio 126
CDMA 22, 23
CDMA2000 3x 28
CDMA2000X 27
CDPD 22
CDR 75
CeBIT 151
cell-ID 35
Cellmania 115
Cellpoint 37
Charles Schwab 133
chatboard 50
Chatting 112
cHTML 14, 42
Cisco 106
CitiBank 119
Clanning 153
Click&Go 154, 176
Clicking: 17 Trends That Drive Your Business—and Your Life 151
Club Nokia 146
CMG 10, 94, 192
CNN 66, 89, 119
Cocooning 161
Colt 106

Commerce One 105
Compact HTML 14
Company Networks 112
Compaq 106, 126
Competition 172, 177
Comverse 59, 94
Confidentiality 17
Content Billing 16
Content Providers 107, 138
core proposition 224
customer 172
Customer Ownership 145
Customer Service 174
Customers 173

D

"daily paper" generation 152
Daimler Chrysler 109
Data integrity 18
DataTAC 22
DDI 118
De Telegraaf 196
Debitel 89
deck 12, 253
DECT 22
Dell 106, 109
Deutsche Telecom 91
DHL Worldwide Express 110, 115
Digital Bridges 115
digital signature 19
Direct Sales 188
Disney 89, 119, 136
Distribution 185
Diversification 222

DLJ 119
Domino 202
Domino Designer 202
Dow Jones 119
Downaging 151
Dutchtone 20

E

E*TRADE 179
E.medi@ 5
Ease of Use 177
Easypay 138
eBay 108
EDGE 28
EDI 105
EFR 87
Egonomics 156
Electrolux 162
email 102, 112
Emerce 193
Encryption 18
Enhanced Full Rate 87
Enhanced Observed Time Difference 35, 36
Entertainment 186
E-OTD 35, 36
EPOC 125
E-procurement 105
Ericsson 4, 7, 13, 32, 88, 94, 118, 125, 193
ERP 105
ETSI 28
European Commission 85
European Law Protecting Privacy 229
European standards 22
Events 13

Excite 119
extranet 77, 112, 189
EZ Access 118
EZ Web 118

F

fad 150
Fantasy Adventures 154
FDMA 22
FedEx 110
Female Think 160
FIEXOS 126
file transfer protocol 102
see FTP
Finphone 210
First-Generation Mobile Networks 22
first-generation systems 22
Flash 62
Flash screen 12
Flex 22
Ford 109
FortuneCity 215
France Telecom 91
Frequency Division Multiple Access 22
FTP 102
Fujitsu 118

G

gateway 6, 240
Gateway Suppliers 128
Gelon 216
General Motors 109

General Packet Radio Services 26, 34
Genie 133
GIF 14
Glenayre 94
Global Crossing 106
Global Positioning System 35, 36
Global System for Mobile Communication 23
GPRS 22, 26, 34, 114
GPS 35, 36
GSM 3, 18, 22, 23, 82

H

Handheld PC 123
Handspring 125
Hardware Suppliers 106
HAVi 41
headset 95
Hello Kitty 50
Hewlett-Packard 106
High Speed Circuit-Switched Data 24
Home Audio/Video interoperability 41
HSCSD 24
HTML 12, 105
HTTP-S 18
Hybrid Devices 51, 124
hype 150
Hypertext Markup Language 12
see HTML

I

I mobile 20
IBM 39, 53, 77, 79, 105

Index

Icon Medialab 141
Icon Toppling 165
ICQ 17, 106, 153
ICT 192
iDEN 22
Identification 17
IDO 13, 118
iMedia 195
i-mode 118
Information 184
information and communication technology 192
Infospace 133
Infrared 124
Ingenico 124
iNotes 201
"In-Store" Competition 178
Integrity 17
Intel 39
Internet 46, 77, 102, 151, 174
    access provider 103
    usage 110
intranet 77, 112, 189
Irrefutability 18
IS-54 23
It's Alive 154

J

Java 15, 119
Jaysar 174
Jini 41
joint marketing systems 141
J-phone 13

K

keys 18
KPN Mobile 5, 89, 92, 143

L

Lernhaut & Hauspie 53
Let's Buy It 174
Levi's 109
Libertel 87
Links 12
Linux 105
Location-based services 38, 144
Location-Dependent 231
Logica 14
Lonely Planet 135
Lotus 201
Lotus Notes 79, 201, 202
Lucent 53
Lycos 119

M

Macromedia 106
Mannesmann 90
Market
    development 222
    penetration 222
Marketing Mix 225
MCI Worldcom 106
mediators 157
Merchants 107, 139
Merita-Nordbanken 137
Message4u 58

MET 19
Microsoft 7, 53, 102, 105, 119
    Exchange 79
    Internet Explorer 62, 105
    Mobile Explorer 7
    Outlook 79
MIDI 14, 50
@info 5, 79, 116, 131, 192
M-ISP 131
Mitsubishi 7, 118
Mobey Forum 138
Mobile
    Access 143
    communication 82
    Company Networks 98
    Electronic Transactions 19
    Explorer 126
    Instant Messaging 17
    Internet 114
    Internet Service Provider 131
    Notes 201, 202
    Operators 86, 129
    Payments 137
    Phone Suppliers 122
    Phone Usage 96
    Portals 133
    Telephone Suppliers 94
    Text Chat 98
    virtual network operator 143
Mobitex 22
Motorola 4, 7, 94, 125, 193
MP3 30, 104, 106, 112
MSN 106

"MTV generation" 152
MVNO 143
m-WorldGate 14
MySAP.com 105

N

NaviRoller™ 13
NEC 118
Netscape 102
Netscape Navigator 62, 105
Network Suppliers 94, 106
New entrants to the market 172, 179
newsgroups 102, 112
Nextel 121
"Nintendo generation" 152
NMT 22
Nokia 4, 7, 39, 88, 94, 118, 125, 193
Nordic Mobile Telephony 22
North American Digital Cellular 23
Northwest Airlines 119
Norwegian Telenor 77
NTT DoCoMo 13, 89, 118, 143

O

objective 222
Omnitel 87
One-2-One 22, 93, 181
Openwave 4, 7
Oracle 105, 129
Orange 87, 88, 91
Origin 77

Index

P

Palm 124
Palm VII 121
Panasonic 5, 94, 118
Payment 57, 175
payment service 110
PC 48
PDA 124
PDC 22, 23
PDC-P 27
Peoplesoft 105
Personal Digital Cellular 23
Personal Identification Number 19
Personalization 228
Personnel 238
Philips 41, 53
Phone.com 5, 128
PHS 22
PIN 19
Pinpoint Networks 115
Pioneer 41, 180
PKI 9
Place 233
Playstation 118, 150
Pocket PC 126
PocketNet 4
Popcorn, Faith 151, 157
portal 106, 233
Porter 172
Positioning 35
prepaid 87
Price 232
    comparison 173
Privacom 54

privacy 156, 229
privacy statement 230
Product 225
    development 222
Promotion 235
Psion 124, 125

Q

QWERTY 48
QXL 154

R

Radati Group 59
Radicchio 137
RealAudio 62
ReFlex 22
Reporting 245
Revenue-sharing 144
Roles in Value Chain 141

S

Sakure Bank 119
Samsung 5, 95
SAP 105
SBC 92
Schwab 179
Scoot 198
search engines 234
Sears 109
Second-Generation Mobile Networks 23
Secure Sockets Layer 74

Security 17
Security layer 11
Service Providers 92, 103
services 115
Session layer 11
SFR 5, 89
Sharp 41
Shell 138
Short Message Service 9, 82
Siebel 79
Siemens 5, 88, 94
Siennax 211
Silicon Graphics 102
Silicon Valley 102
SIM 18, 66
SIM Application Toolkit 19
SIM card 21, 74, 126
SingTel 93
Sky Go 140
Sky Melody 50
Sky Radio 212
Small Indulgences 163
smart agents 157
Smart phone 123
Smarttone 26
SMS 9, 82
SMS traffic 96
Soft buttons 12
Software Suppliers 105
Software Tools Suppliers 128
Sonera 89, 137
Sony 7, 41
spam 229
Speech Technology 38
Sprint 106, 121

SSL 74
Standards 42
STK 19
Subscriber Identity Module 18, 66
Substitutes 172, 183
Sun Microsystems 41, 119
Suppliers 172, 188
Symbian 43, 125
Symbol 126
Synchronization 38

T

T9 17, 49
TACS 22
tags 12
Talkline 89
Tamagotchi 119
Tare Panda 50
TDMA 22, 23
TDOA 35, 36
Tegic 17, 49
Tele Danmark 89
Telecom Italy 91
Telefonica 91, 106
Telfort 20
Telstra 26, 126
Terenci 133
Terminals 114
TETRA 22
Third-Generation Mobile Networks 28
    see 3G
Time Difference of Arrival 35, 36
Time Division Multiple Access 23
Time of Arrival 35, 36

Index

TLS 74
T-Mobil 126
TOA 35, 36
Toshiba 39
Total Access Communications System 22
Toys 'R Us 108, 181
TPG 110
Transaction layer 11
Transaction Providers 137
Transactions 144, 186
Transport layer 11
Transport Layer Security 74
Travelocity 133
trend 150
Triangulation 36
tucows 106
Twigger 214

U

UAProf 9
UMTS 22, 28, 29
UMTS Auctions 85
Unified Messaging 58
Uniform Resource Locators (URLs) 8
Unisys 94
United Customers 174
Universal 119
Universal Plug and Play 41
Unstructured Supplementary Services Data 21
UPS 110
USSD 21
UUnet 103

V

value-added services platforms 94
Vending Machines 188
Verifone 124
Vigilante Consumer 159
Virgin Mobile 21, 92, 93, 136
virtual mobile operator 21
virtual network operators 92
virtual private network 99
    see VPN
Visor 125
Vivendi 90, 133
Vizzavi 89, 133, 143
Vodafone 89, 90, 126, 133
Voice Recognition 53
Voice Xpress 53
VoiceXML 39
VPN 99

W

Walkman 52
Walled Garden 145
Walmart 181
WAP 2
    architecture 7
    rowsers 7
    client 7
    gateway 8
    home-page builders 253
    Lock 145
    protocol 11
    Service Broker 193

site 253
toolkit 253
WAP Forum 4, 18, 126, 193
WAPDrive 215
WAPit 96
Wapportal 136
W-CDMA 29
Webraska 128
Websphere Transcoding Publisher 79
Wehkamp 108
Wideband CDMA 28
Wideband CDMAOne 27
Wideband Code Division Multiple Access 29
Wildfire 94
WIM 18, 75
Wireless Application Environment 11
Wireless Application Protocol 2
wireless bitmaps 254
Wireless Datagram Protocol 11
Wireless Identity Module 18, 75
Wireless Markup Language 12
Wireless Session Protocol 11
Wireless Telephony Application 15, 72
Wireless Telephony Application Interface 15
Wireless Transaction Protocol 11
Wireless Transport Layer Security 11, 18, 74
WML 15
WML Script 15
World Wide Web 46, 102
World Wide Web Consortium 39
WSP 11
WTA 15, 72
WTAI 15
WTLS 18, 74, 78

X

Xoip 58
XS4ALL 131, 218

Y

Yahoo! 17, 66, 89, 106, 119, 146, 180

About the Authors

Ingrid Vos and Pieter de Klein were jointly responsible for the successful introduction of @info, the world's first commercial WAP portal. @info and other mobile portal initiatives have been aggressively marketed in combination with Nokia's first WAP-enabled handset throughout Europe, where M-business is widespread. Both authors have worked for KPN Mobile (Netherlands), an innovative mobile operator doing business in several countries.

Ingrid Vos has broad experience in marketing corporate telecommunication solutions. She is manager of New Business and Strategy at KPN Mobile, responsible for creating marketing propositions like m-commerce, location-based services, and mobile office applications. KPN Mobile is the leading mobile operator in the Netherlands and has presence in several other countries.

Pieter de Klein, formerly a marketer with KPN Mobile, now heads the Marketing & Sales department of PrivaCom. PrivaCom is a service innovator on the cutting edge of telecoms and internet, developing services like Call me now! and SIM Callback. The latter was awarded the GSM association innovation award 2001.

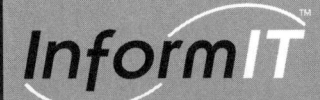

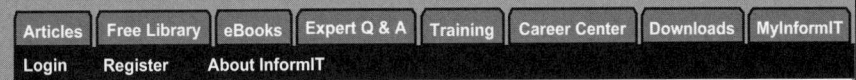

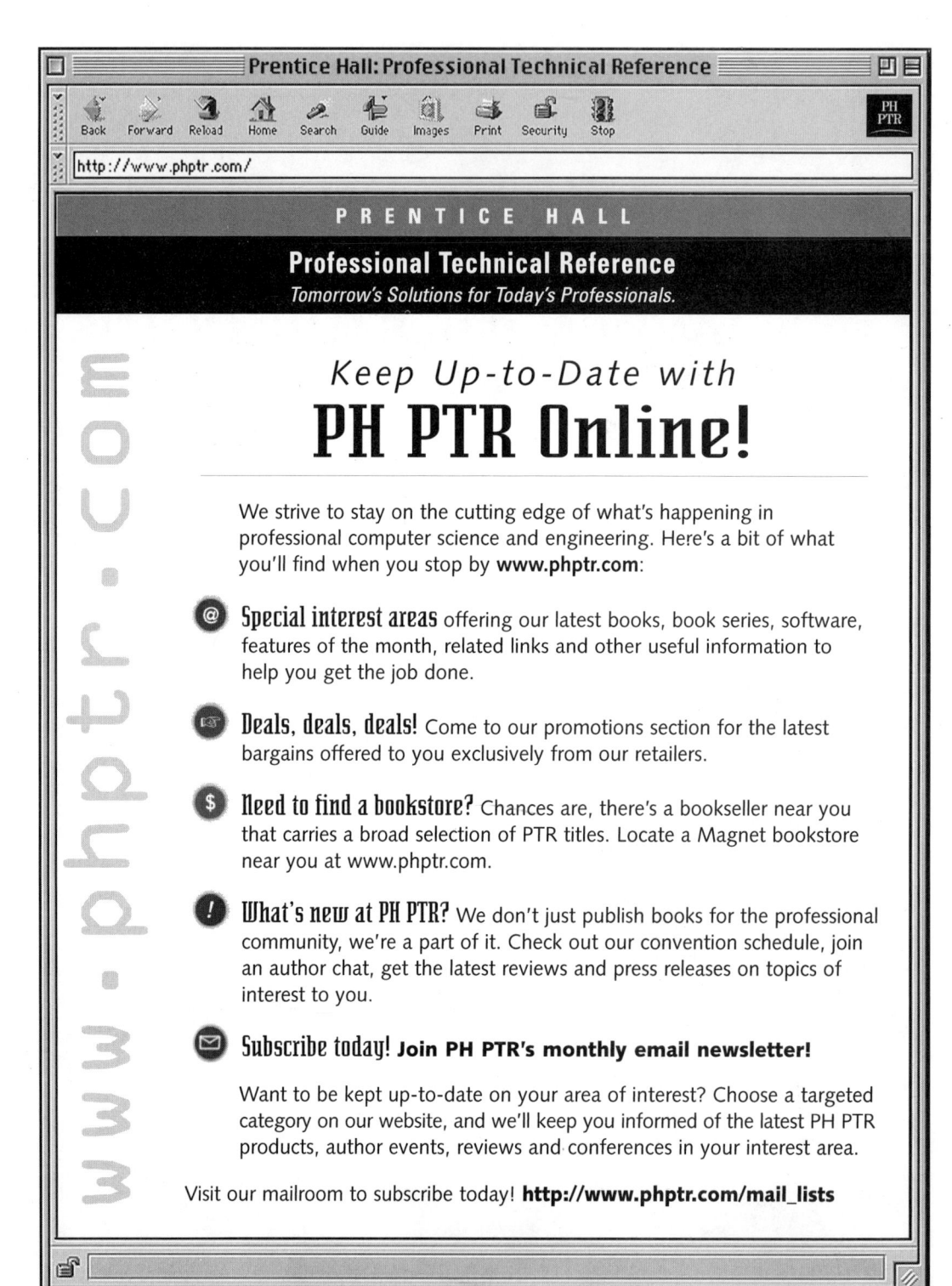

PRENTICE HALL
Professional Technical Reference
Tomorrow's Solutions for Today's Professionals.

Keep Up-to-Date with
PH PTR Online!

We strive to stay on the cutting edge of what's happening in professional computer science and engineering. Here's a bit of what you'll find when you stop by **www.phptr.com**:

@ Special interest areas offering our latest books, book series, software, features of the month, related links and other useful information to help you get the job done.

Deals, deals, deals! Come to our promotions section for the latest bargains offered to you exclusively from our retailers.

$ Need to find a bookstore? Chances are, there's a bookseller near you that carries a broad selection of PTR titles. Locate a Magnet bookstore near you at www.phptr.com.

! What's new at PH PTR? We don't just publish books for the professional community, we're a part of it. Check out our convention schedule, join an author chat, get the latest reviews and press releases on topics of interest to you.

Subscribe today! Join PH PTR's monthly email newsletter!

Want to be kept up-to-date on your area of interest? Choose a targeted category on our website, and we'll keep you informed of the latest PH PTR products, author events, reviews and conferences in your interest area.

Visit our mailroom to subscribe today! **http://www.phptr.com/mail_lists**